Klarissa Niedermeier

Numerical investigation of a thermal storage system using sodium as heat transfer fluid

Karlsruhe Institute of Technology
KIT SCIENTIFIC REPORTS 7755

Numerical investigation of a thermal storage system using sodium as heat transfer fluid

by
Klarissa Niedermeier

KIT Scientific Publishing

Report-Nr. KIT-SR 7755

Dissertation, Karlsruher Institut für Technologie
KIT-Fakultät für Chemieingenieurwesen und Verfahrenstechnik

Tag der mündlichen Prüfung: 5. April 2019
Referent: Prof. Dr.-Ing. Thomas Wetzel
Korreferent: Prof. Dr.-Ing. Thomas Schulenberg

Impressum

KIT Scientific Publishing

Karlsruher Institut für Technologie (KIT)
KIT Scientific Publishing
Straße am Forum 2
D-76131 Karlsruhe

KIT Scientific Publishing is a registered trademark
of Karlsruhe Institute of Technology.
Reprint using the book cover is not allowed.

www.ksp.kit.edu

Print on Demand 2019 – Gedruckt auf FSC-zertifiziertem Papier

ISSN 1869-9669
ISBN 978-3-7315-0930-1
DOI 10.5445/KSP/1000093902

Numerical investigation of a thermal storage system using sodium as heat transfer fluid

Zur Erlangung des akademischen Grades
Doktor der Ingenieurwissenschaften (Dr.-Ing.)
der KIT-Fakultät für Chemieingenieurwesen und
Verfahrenstechnik des
Karlsruher Institut für Technologie (KIT)

genehmigte
Dissertation

von

Dipl.-Ing. Klarissa Niedermeier
aus Mudau

Tag der mündlichen Prüfung:	05. April 2019
Referent:	Prof. Dr.-Ing. Thomas Wetzel
Korreferent:	Prof. Dr.-Ing. Thomas Schulenberg

Vorwort

Die vorliegende Arbeit entstand während meiner Tätigkeit in der Arbeitsgruppe Karlsruher Flüssigmetalllabor am Institut für Kern- und Energietechnik vom Oktober 2014 bis September 2018.

An dieser Stelle möchte ich allen danken, die mich beim Entstehen dieser Arbeit unterstützt haben. Mein besonderer Dank geht an Prof. Thomas Wetzel für die fachliche Betreuung meiner Arbeit und das entgegengebrachte Vertrauen. Herrn Prof. Thomas Schulenberg danke ich für die Übernahme des Korreferats. Außerdem möchte ich mich bei Dr. Luca Marocco bedanken für die fachliche Unterstützung bei der Durchführung der Simulationen. Die tolle Zusammenarbeit war essentiell für das Gelingen dieser Arbeit.

Desweiteren gilt mein Dank meinen Kollegen am Karlsruher Flüssigmetalllabor. Dabei danke ich vor allem Jonathan Flesch, der immer ein offenes Ohr hatte und durch konstruktive Diskussionen die Inhalte der vorliegenden Arbeit vorangebracht hat. Julio Pacio danke ich für den fachlichen Austausch und das Korrekturlesen. Auch den Kollegen der Arbeitsgruppe Wärme- und Stoffübertragung am Institut für Thermische Verfahrentechnik möchte ich für die fachliche und persönliche Unterstützung danken.

Außerdem danke ich Martin Pfuhl, Judith Kleinheins, Niccola Zancan und Manuel Meßmer, deren hervorragende und sorgfältige studentische Arbeiten eine große Hilfe bei der Erstellung dieser Arbeit war.

Besonderer Dank gilt Mary Schäfer für die sprachlichen Korrekturen an dieser Arbeit. Auch an Silvan Niedermeier geht ein Dankeschön für seine sprachlichen Anmerkungen.

Weiterhin danke ich meinen Freunden und meinem Partner Thomas für die Unterstützung während der Promotion und des Studiums. Abschließend bedanke ich mich bei meiner Familie, die mich mein ganzes Leben und auch auf dem Weg zu dieser Arbeit stets bestärkt hat.

Karlsruhe, im April 2019 *Klarissa Niedermeier*

Kurzfassung

Konzentrierende Solarthermieanlagen stellen durch die Verwendung thermischer Speicher elektrische Leistung zuverlässig auch nach Sonnenuntergang bereit. Solarturmkraftwerke, die sich durch hohe Konzentrationsverhältnisse auszeichnen, nutzen momentan eine Salzschmelze, sogenanntes Solarsalz, welches sowohl als Wärmeträgerfluid im thermischen Receiver als auch als Speichermedium eingesetzt wird. Allerdings sind weitere Kostenreduktionen notwendig, damit sich die konzentrierende Solarthermie neben der Photovoltaik durchsetzen kann. Dafür werden seit einigen Jahren alternative Fluide, die auch bei Temperaturen oberhalb 600 °C arbeiten können, getestet, um eine Erhöhung des Gesamtwirkungsgrads der Anlage und eine Reduzierung der Stromgestehungskosten zu erzielen.

Flüssigmetalle eignen sich dafür besonders aufgrund ihrer hervorragenden Wärmeübertragungseigenschaften und des breiten Temperaturbereichs, in dem sie eingesetzt werden können. Dabei wird vor allem Natrium untersucht, das sich durch eine sehr hohe Wärmeleitfähigkeit und eine niedrige Schmelztemperatur auszeichnet. Zwei konzentrierende Solarthermieanlagen im Pilotmaßstab haben die Eignung von Natrium als Wärmeträgerfluid bereits gezeigt, in den 1980ern in Almería, Spanien, und seit 2017 in Jemalong, Australien. Das erhitzte Natrium wurde in diesen Anlagen direkt in zwei Tanks gespeichert, wie es auch in Anlagen für Solarsalz der Fall ist, obwohl sich Natrium durch seine vergleichsweise niedrige Energiedichte nur bedingt für die Nutzung als Speichermedium für diese Konfiguration eignet. Alternative Speichermöglichkeiten mit Natrium werden zwar in der Literatur vorgeschlagen, jedoch wurde keine systematische Evaluierung der möglichen Speicheroptionen mit Natrium als Wärmeträgerfluid durchgeführt.

Die vorliegende Arbeit schließt diese Forschungslücke. Die durchgeführte Evaluierung führte zu dem Ergebnis, dass folgende drei Speicherkonfigurationen in Kombination mit Natrium als Wärmeträgerfluid vielversprechend sind: sensible Schichtenspeicher mit einem Festbett, Latentspeicher mit Phasenwechselmaterialien und thermochemische Speicher, die sich durch hohe Speicherdichten auszeichnen. Von diesen wurde der Schichtenspeicher detailliert untersucht. Im Ladefall durchströmt heißes Natrium diesen Speicher von oben her, wobei es seine thermische Energie an die Füllkörper abgibt, im Entladefall durchströmt kaltes Natrium den Speicher von unten her und nimmt die Energie auf. Zwischen der heißen und kalten Phase bildet sich im Tank eine Dichte- und Temperatursprungschicht, die sogenannte Thermokline. Je kleiner diese Thermokline, desto besser ist die Effizienz des Speichers.

Der Temperaturverlauf in einem solchen Speicher wurde berechnet, indem die Energieerhaltungsgleichungen sowohl zwei-dimensional als auch zwei-phasig mit Berücksichtigung der Wärmeleitung in kugelförmigen angenommenen Füllkörpern unter Verwendung der Finite-Volumen Methode gelöst wurde. Die Validierung des Codes erfolgte mit Daten aus Experimenten mit Salzschmelzen und Öl. Die Simulationen in der vorliegenden Untersuchung zeigten, dass die Verwendung von Natrium, aufgrund der exzellenten Wärmeübertragung zwischen Flüssigmetall und Füllkörpern, zu hervorragenden Effizienzen im zyklischen Betrieb (Laden/Entladen) führt. Im Stillstand wächst die Thermokline jedoch durch die hohe thermische Diffusivität von Natrium schnell an und die Effizienz beim anschließenden Entladen sinkt. Dies zeigte sich auch im Vergleich mit Solarsalz und Hochtemperatursalzen, die im zyklischen Betrieb im Festbett durch geringere Wärmeübertragungskoeffizienten zwischen Flüssigkeit und Festbett niedrigere Effizienzen erzielen, jedoch im Stillstand durch die niedrigere thermische Diffusivität Vorteile aufweisen. Die durchgeführte Parameterstudie zeigte, dass eine niedrige Porosität, ein geringes Verhältnis von Tankdurchmesser zu -höhe und kleine Füllkörper sowohl das zyklische als auch das Stillstandsverhalten begünstigen.

Eine serielle Multi-Tank Anordnung ermöglichte eine weitere Optimierung. Durch diese Konfiguration können kleinere Verhältnisse von Tankdurchmesser zu -höhe erreicht werden verglichen mit einem einzelnen Tank. Außerdem wird die axiale Wärmeleitung durch die thermische Isolierung der einzelnen Tanks voneinander unterbrochen. Vor allem bei längeren Stillstandszeiten zeigte sich der Vorteil dieser Anordnung, da nur diejenigen Tanks sich thermisch ausgleichen, in denen sich die Thermokline befindet. Die übrigen Tanks erfahren keine Temperaturänderung, unter der Annahme, dass diese thermisch ausreichend isoliert sind.

Basierend auf einem Vergleich verschiedener Speicheroptionen konnte in dieser Arbeit ein vielversprechendes Speichersystem für Natrium identifiziert werden. Der Schichtenspeicher überzeugt vor allem durch geringe Kosten und eine hohe Speicherdichte durch die Zugabe von festem Füllmaterial. Weiterhin konnte mithilfe numerischer Simulationen gezeigt werden, dass sich ein solcher Speicher erfolgreich zyklisch betreiben lässt und sogar höhere Effizienzen als mit Salzschmelzen in Festbettspeichern erreicht werden können. Darüber hinaus ließ sich der thermische Ausgleich im Festbett in längeren Stillstandzeiten durch die gezeigte Multi-Tank Anordnung verbessern.

Abstract

Concentrating solar power plants use low-cost thermal energy storage systems to provide electricity reliably and on demand. The state-of-the-art central receiver systems use a two-tank storage solution with a molten salt mixture as heat transfer and storage medium. However, as further cost reductions are necessary to push the concentrating solar power technology, higher operating temperatures (above 600°C) are aimed for to increase the overall efficiency of the power plant and decrease the levelized cost of electricity. Solar salt, currently in use, cannot be employed at this temperature level owing to its decomposition temperature. As alternative heat transfer fluids, liquid metals are proposed to be used in the thermal receiver.

They qualify due to their excellent heat transfer properties and large operating temperature range. Especially sodium is attractive due to its outstanding thermal conductivity and low melting temperature. Pilot concentrating solar power plants with sodium have already been tested in Almeria, Spain, in the 1980s and in Jemalong, Australia, since 2017, showing high receiver efficiencies. The heated sodium is stored using a two-tank system as usual for molten salt, although sodium is not advantageous as a storage medium in this kind of system due to its relatively low energy density. Even though sporadic storage solutions have been proposed in the recent past, a thorough evaluation of storage systems for sodium as a heat transfer fluid has not yet been performed.

Therefore, in the present study a systematic evaluation of a range of thermal energy storage systems and configurations has been conducted, showing that three kinds of systems are advantageous when using sodium as a heat transfer fluid. These three systems are: sensible storage with a packed bed, latent storage using encapsulated phase change materials and thermochemical storage with large storage densities. The sensible storage using a packed bed with liquid sodium flowing through when discharged and charged has been chosen for a detailed investigation in this work. This packed bed is charged with hot sodium flowing from the top down and discharged with cold sodium flowing from the bottom up. Thus, a so-called thermocline zone builds up in the region where cold fluid comes in contact with hot fluid. The better the separation and hence the smaller the thermocline zone, the more efficiently can heat be stored in this kind of storage system.

For a performance analysis of the chosen thermal energy storage system a two-dimensional two-phase transient model has been used, including also the radial thermal conduction in the spherical filler particles. The differential equations were solved using the finite volume method and the simulation was conducted in MATLAB. The code was validated with data from molten salt and oil experiments. The results showed that the use of sodium leads to high discharge efficiencies during cyclic operation due to the excellent heat transfer between the liquid metal and the solid particles. However, during standby the thermocline zone degrades fast due to the large thermal diffusivity of liquid sodium. A parametric study performed in the present work shows that a low bed porosity, a small tank diameter-to-height ratio, and small filler particles lead to an improved behaviour both during discharge and standby. In comparison with solar salt at state-of-the-art temperatures and advanced salts at elevated temperatures, higher discharge efficiencies are gained for sodium in the selected reference case. However, during standby, a reduced thermocline degradation shows the advantage of salts compared with sodium due to their lower thermal diffusivity.

In order to further improve performance in the selected storage system with liquid sodium, a multi-tank configuration was analysed, where multiple tanks are operated in series. This arrangement makes it possible to realize smaller tank diameter-to-height ratios and, additionally, reduces the axial conduction due to thermal insulation of the tanks compared with a single-tank arrangement. The results showed that both the discharge efficiency and especially the long-term standby behaviour improved, as now only a part of the overall storage volume, where the thermocline zone is present, is affected by thermal conduction. The remaining tanks keep their temperature if well-insulated.

In summary, this study selected a promising packed-bed thermal energy storage system for further investigation based on an evaluation of various suitable storage options, particularly in view of the reduced cost and increased storage density. Furthermore, a numerical model was developed to determine the temperature distributions during cyclic operation and standby. The simulation results showed that a packed-bed thermal energy storage system can not only be operated with liquid sodium but is even advantageous during discharge when compared with molten salts used in packed beds. However, considering long standby periods, technical solutions need to be found to minimize the thermocline degradation. As one possibility, a multi-tank configuration was proposed and simulated in this work.

Table of Contents

Nomenclature

Roman letters

A	Surface area	m^2
a	Thermal diffusivity	m^2/s
C	Specific storage material cost	€/kg
c_p	Specific heat capacity	J/(kgK)
D	Tank diameter	m
d	Filler diameter	m
d_{32}	Sauter diameter	m
g	Gravity	m/s^2
H	Tank height	m
h	(Mass specific) enthalpy	J/kg
h_{v}	Volumetric heat transfer coefficient	$W/(m^3K)$
h_{w}	Heat transfer coefficient at the tank wall	$W/(m^2K)$
$h_{\mathrm{w,v}}$	Volumetric heat transfer coefficient at the tank wall	$W/(m^3K)$
i	Space index variable	–
j	Time index variable	–
k	Effective thermal conductivity	W/(mK)
l	Flow length	m
L_c	Characteristic length	m
M	Storage material cost	€/(kWh)
m	Mass	kg
$\dot{m}$	Mass flow	kg/s
Δp	Pressure difference	Pa

P	(Pumping/electrical/thermal) power	W
Q	Energy, storage capacity	J,MWh
R	Tank radius	m
r	Coordinate along tank radius	m
s	Thickness of wall and insulation layers	m
s_{v}	Specific surface	1/m
T	Temperature	°C,K
t	Time	s
u	Velocity in packed bed	m/s
u_0	Velocity through empty cross-sectional area	m/s
V	Volume	m^3
$\dot{V}$	Volume flow	m^3/s
x	Coordinate along tank height	m
y	Coordinate along particle radius	m

Greek letters

α	Heat transfer coefficient	$W/(m^2K)$
β	Interpolation factor spatial discretization	–
β	Volume expansion coefficient	1/°C
ε	Porosity	–
ζ	Thermocline zone related to tank height	–
η	Discharge efficiency	–
λ	Thermal conductivity	W/(mK)
μ	Dynamic viscosity	Pas
ν	Kinematic viscosity	m^2/s
ϕ	Polar angle	rad
φ	Azimuth	rad
ρ	Density	kg/m^3
Θ	Time scheme factor	–

Acronyms

CFD	Computational fluid dynamics
CSP	Concentrated solar power
CV	Control volume
D	Dimension
FV	Finite volume
HTS	High-temperature salt
HX	Heat exchanger
LBE	Lead-bismuth eutectic
LCOE	Levelized cost of electricity
LM	Liquid metal
P	Phase
PCM	Phase-change material
PV	Photovoltaics
TES	Thermal energy storage

Non-dimensional parameters

Bi	Biot number: $\mathrm{Bi} = \alpha L_\mathrm{c}/\lambda_\mathrm{s}$
Fo	Fourier number: $\mathrm{Fo} = \tau = at/L_\mathrm{c}^2$
Gr	Graetz number: $\mathrm{Gr} = gL_c^3\beta\Delta T/\nu^2$
Nu	Nusselt number: $\mathrm{Nu} = \alpha d/\lambda_\mathrm{f}$
Pe	Peclet number: $\mathrm{Pe} = \mathrm{Pr} \cdot \mathrm{Re_p}$
Pr	Prandtl number: $\mathrm{Pr} = \mu_\mathrm{f} c_{p\mathrm{f}}/\lambda_\mathrm{f}$
Ra	Rayleigh number: $\mathrm{Ra} = \mathrm{Gr} \cdot \mathrm{Pr} = gL_c^3\beta\Delta T/(\nu a)$
$\mathrm{Re_p}$	Reynolds number: $\mathrm{Re_p} = u_0 d\rho_\mathrm{f}/\mu_\mathrm{f}$
τ	Time: $\tau = \mathrm{Fo}$
θ	Temperature: $\theta = \frac{T(t)-T_\infty}{T(t=0)-T_\infty}$
R	Radial distance tank: $R = r/(D/2)$

X	Axial distance tank: $X = x/H$
Y	Radial distance particle: $Y = y/(d/2)$

Subscripts

∞	Infinite
0	Ambient
bed	Packed bed
c	Characteristic
chg	Charge
dis	Discharge
e,E	East
eff	Effective
f	Fluid
fr	Fluid, radial direction
fx	Fluid, axial direction
in	Inlet
m	Mass specific
max	Maximum
melt	Melting
min	Minimum
mix	Mixed/homogeneous
n,N	North
P	Centre of node
p	Particle
r	Residence
reac	Reaction
s	Solid
s,S	South
sens	Sensible

stby	Standby
thcl	Thermocline
tot	Total
v	Volume specific
w	Tank wall
w,W	West

1 Introduction

In order to meet carbon dioxide reduction targets, the share of renewables has increased worldwide. Within the renewable energy technologies the concentrating solar power (CSP) is of particular interest due to the applicability of a low-cost thermal energy storage system. This technology uses concentrated solar energy to heat up a transport medium which transfers the heat to a power cycle, where electricity is produced (Figure 1.1).

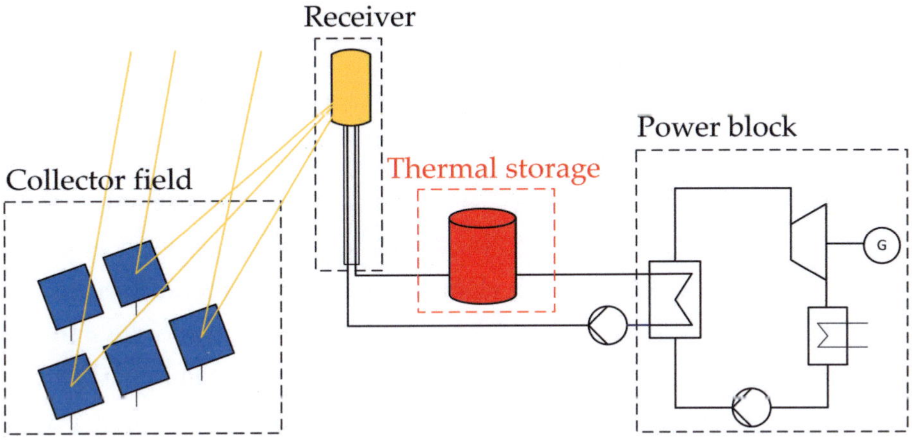

Figure 1.1: Scheme of solar tower plant with four subsystems: collector field, receiver, thermal storage and power block

Owing to its integrated thermal energy storage systems, CSP technology is able to react on demand, and therefore to stabilize fluctuations in a grid with a high share of renewables. The storage system is a key component of the CSP technology, as it allows it to be competitive with photovoltaics coupled with battery storage. Although the commercial use of CSP began only in 2007, with exception of the Solar Energy Generating Systems (SEGS) parabolic trough plants in California in the 1980s, CSP had already reached a capacity of 5 GW in 2016 [1], which was 0.08% of the global total electricity capacity [2].

1.1 Motivation

To further increase the share of CSP, the levelized costs of electricity of these systems need to be decreased [3]. For this reason, research is looking at alternatives to the currently used heat transfer fluid "solar salt", ones with higher upper temperature limits (> 600 °C) leading to higher overall efficiencies [4]. For these elevated temperatures, liquid metals are possible candidates [5, 6]. Among the liquid metals, sodium is the prime option owing to its low melting temperature and high thermal conductivity ($\approx$ 100 higher than solar salt). Simulations have already shown that the replacement of solar salt with sodium in a state-of-the-art CSP plant results in a reduction of the levelized costs of electricity of 16% even without increasing the upper temperature [7]. The high thermal conductivity allows a higher heat transfer rate at the same temperature difference between wall and bulk temperature; thus, the receiver area can be reduced and material costs are lower. Additionally, the smaller receivers lead to reduced heat losses and therefore to higher receiver efficiencies. For upper temperatures of above 600 °C in the central receiver, which could be reached with sodium as the heat transfer fluid, the state-of-the-art solar salt storage system is no longer applicable. However, thermal energy storage systems for these temperatures and coupled with liquid metals as heat transfer fluids have scarcely been investigated to date.

1.2 Current state of research

Currently operational CSP plants with solar salt as the heat transfer fluid use a direct two-tank configuration as thermal energy storage. This configuration was also used in the experiments conducted in the IEA-SSPS facility with an electric output of 500 kW in Almería, Spain, in the 1980s. In this plant, sodium has proven to be a highly efficient heat transfer fluid for temperatures of up to 530 °C [8]. The storage time was 2 h. Recently, the use of sodium in CSP plants was re-demonstrated in a pilot plant in Jemalong, Australia, with temperatures of up to 565 °C [9]. This pilot plant with an electric output of $\approx$ 1 MW is also equipped with a direct two-tank storage for 3 h. This direct two-tank configuration requires a large quantity of sodium (especially with higher storage times), meaning high costs at relatively low storage density. Additionally, the large mass of sodium entails an increased potential risk of a sodium fire.

In the literature, sporadic studies exist regarding thermal energy storage systems for liquid metal at the aimed temperature. Pomeroy (1979) [10] proposed a packed bed of iron spheres as storage medium in direct contact with sodium. No further experimental and numerical investigation have yet been performed, to the best of the author's knowledge. Recently, Kotzé (2014) [11] investigated a latent thermal energy storage with $AlSi_{12}$ and sodium-potassium as the heat transfer fluid to achieve higher storage densities at higher upper temperatures of 620 °C. However, the corrosive nature of aluminium led to complications and limited experimental results. Wilk (2016) [12] examined the costs of a liquid tin receiver loop with an Al-Si alloy as a combined sensible and latent storage medium with temperatures of up to 1350 °C. Demonstration of this loop is planned, but no results have yet been published. De la Calle et al. (2017) [13] published a system analysis of a central receiver with sodium, a latent heat storage and a CO_2 power cycle with a resulting reduction of the levelized cost of electricity when compared with a conventional molten salt system.

Another approach is to use a direct contact between the latent heat material and sodium, as shown by Kee et al. (2018) [14].

All in all, only the direct two-tank storage solution has been experimentally tested for liquid metals to date. Only few alternative storage solutions are proposed and no systematic analysis of different storage options suitable for liquid metals for upper temperatures > 600 °C has yet been done. Such an analysis should include an evaluation of thermal energy storage options based on technological and basic economic criteria and a detailed theoretical design analysis of promising candidate systems.

1.3 Research task

The aim of this doctoral thesis is to propose a suitable storage option for a liquid metal CSP system for upper temperatures > 600 °C. In order to advance the current state of research, the work has two main objectives:

- In a first step, it will be determined whether thermal energy storage for sodium as the heat transfer fluid for temperatures > 600 °C can be realized. This evaluation will be based on a theoretical technical and simplified economical analysis. A storage system type will be selected in a systematic evaluation of thermal energy storage systems beyond the currently used two-tank storage, taking also latent and thermochemical systems into consideration, which are promising due to their high storage densities.
- In the second step, based on this evaluation, a storage system type most suited for a CSP system with liquid metal as the heat transfer fluid will be theoretically investigated and optimized at a component level. For this purpose, a numerical model will be established and validated based on the current state of research. In a parametric study, the optimal parameters will be identified and a comparison with state-of-the-art fluids will be conducted. Based on this work it will be possible to determine the important parameters for a test on lab/pilot scale.

A thermo-hydraulic and economic analysis at system level with an integration in a CSP plant is beyond the scope of this work. The focus is placed on investigating the advantages and limitations of using sodium in the selected storage system and not on determining the overall CSP plant efficiency and levelized costs of electricity, though results might be used in such analyses as the one performed in the related PhD work of Fritsch [7].The outline of the present thesis is shown in Figure 1.2.

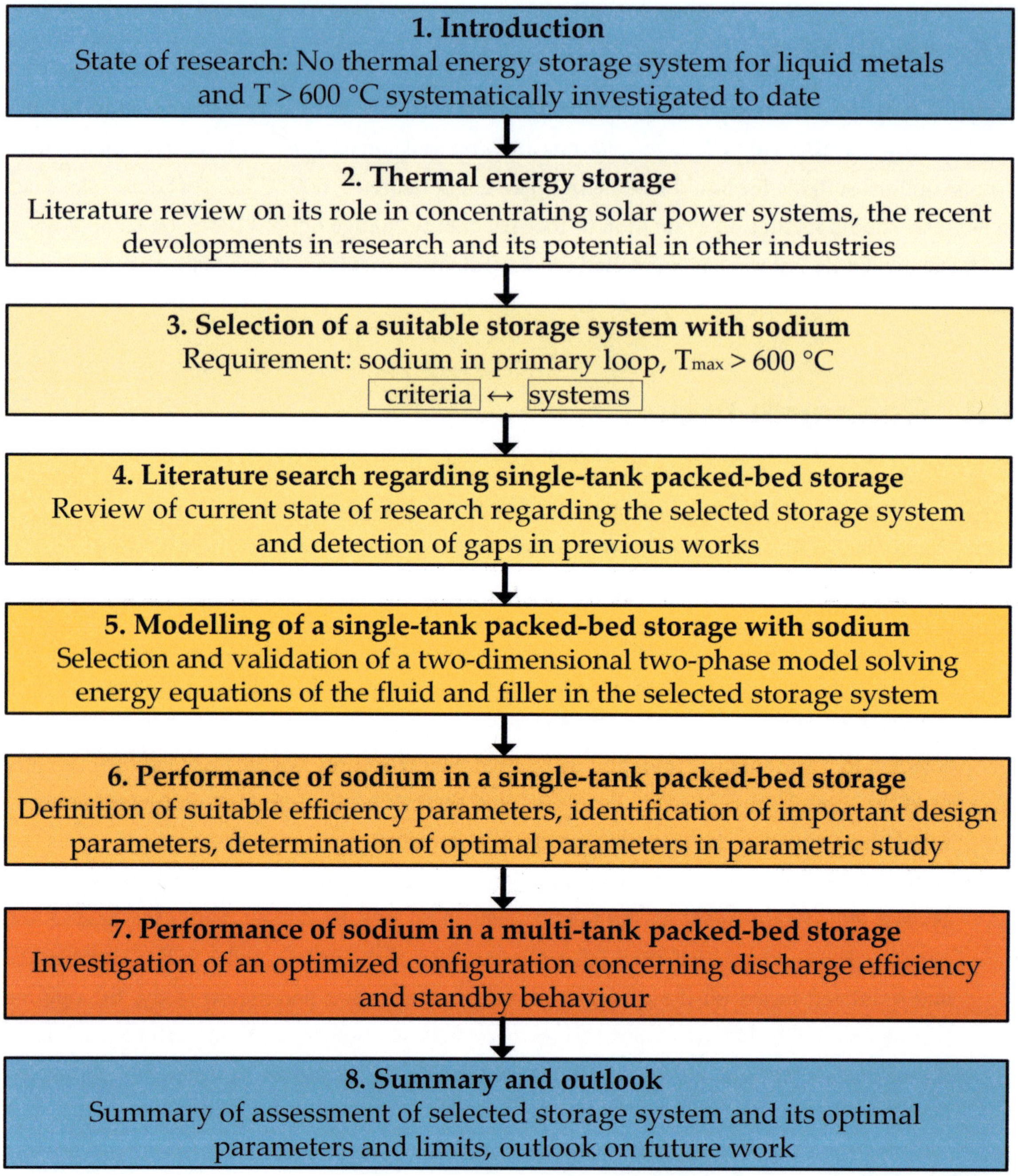

Figure 1.2: Outline of this doctoral thesis

2 Thermal energy storage

Thermal energy storage systems are used in a wide range of applications, depending on their size and temperature level, from household hot-water heaters with storage to large-scale seasonal storage [15]. In this work, the focus is on the use of thermal energy storage systems in CSP plants. Their role for the CSP industry is explained in Section 2.1 including the state-of-the-art storage configuration, the competition of CSP with PV and the cost-reduction goals of CSP. Furthermore, recent research regarding the reduction of the costs of thermal energy storage is presented in Section 2.2. Finally, the potential of integrating high-temperature thermal energy storage in energy-intensive industries is examined in Section 2.3.

2.1 Role in CSP technology

Depending on demand, site conditions and the type of solar thermal system, a storage system can:

- Supply power after sunset and therefore increase the plant's capacity factor (see below)
- Guarantee output at any time and therefore provide electricity reliably to the grid
- Buffer cloud periods
- Prevent freezing
- Enable pre-heating of components

The capacity factor is the ratio of annual electricity full load output and the rated power multiplied with the maximum hours of the year, in other words, the ratio of full load hours to actual hours a year. It can be understood as a degree of utilization of the plant [16].

Figure 2.1 shows an exemplary incoming direct normal irradiation distribution (*dark blue*) during the course of a day. The thermal energy storage system makes it possible to store the heat during start-up (*light blue*) in the morning and to release it after sunset (*light green*) in order to provide a constant electricity output (*orange*).

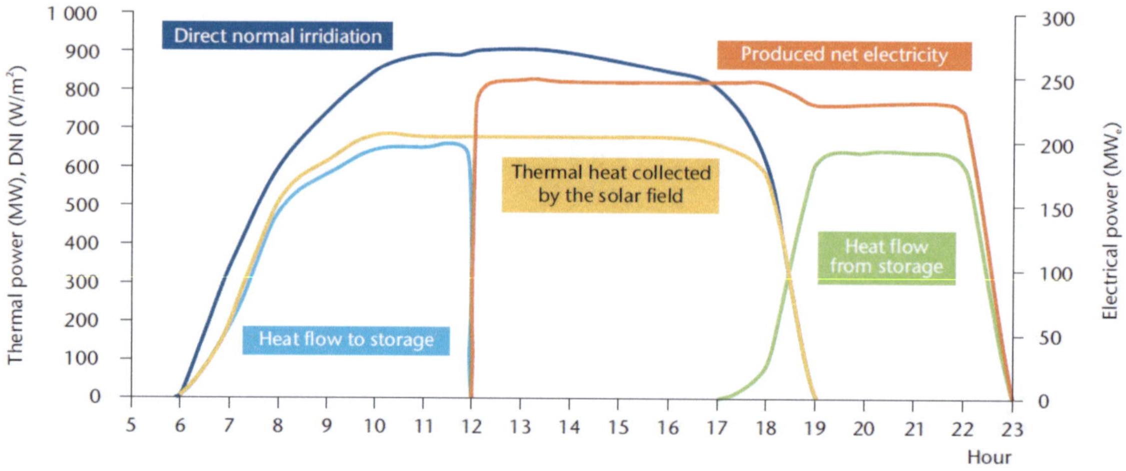

Figure 2.1: Schematic electricity production shift to cover evening peaks (from IEA Technology Roadmap Solar Thermal Electricity [17])

2.1.1 State-of-the-art

The first thermal energy storage system investigated in the 1980s for research purposes as part of the demonstration plant Solar One was a single-tank storage with a packed bed of rocks and sand and thermal oil as the heat transfer fluid at temperatures of up to 305 °C and a storage capacity of 170 MWh [18]. In 1995, the plant was enlarged by adding heliostats and named Solar Two. The heat transfer and storage fluid was changed to molten salt, and a 105-MWh two-tank arrangement was tested for the first time for temperatures of up to 565 °C, which is now the state-of-the-art storage system [19]. The first commercial storage system was part of the parabolic trough plant SEGS I in the 1980s, with mineral oil as heat transfer and storage fluid. It was a two-tank system with a 120-MWh capacity [20]. Herrmann et al. [20] list further storage tests in research plants (Table 2.1).

In currently operational CSP plants a two-tank configuration is used as the thermal energy storage system, with one tank containing cold fluid and one containing the heated fluid. It is used, for example, in the CSP plant Gemasolar, which is located in Andalusia, Spain (Figure 2.2). Gemasolar is a central receiver plant with 2650 heliostat mirrors that reflect the sunlight to the top of a 140-m-high tower to the central receiver. In the receiver, 8500 tons of solar salt are heated up during the day from 290 °C to 565 °C and a thermal power of 120 MW is produced. The hot salt is stored in the hot tank, whereas the cold storage tank is drained during the day. The thermal storage can be discharged for 15 h, and thus enable a 24-h operation of the plant. The heated salt is transferred to a steam generator, and the heated steam expands in the turbine. The plant delivers an electric power of 19.9 MW to the grid [21].

Figure 2.3 shows currently operational central receiver systems with and without integrated thermal energy storage according to their electric power output and their start of operation. Although the addition of a storage system leads to increased investment costs and therefore LCOE,

Table 2.1: Tested thermal energy storage systems for CSP in the past [20]

Project	Storage medium	T_{max}	Storage type	Thermal Capacity
Irrigation pump, AZ	Oil	228 °C	Single-tank	3 MWh
IEA-SSPS, Spain	Oil	295 °C	Single-tank	5 MWh
SEGS I, CA	Oil	307 °C	Two-tank	120 MWh
IEA,SSPS, Spain	Oil/Iron	295 °C	Single-tank	4 MWh
Solar One, CA	Oil/Sand/ Rock	304 °C	Single-tank	170 MWh
CESA-1, Spain	Salt	340 °C	Two-tank	12 MWh
THEMIS, France	Salt	450 °C	Two-tank	40 MWh
Solar Two, CA	Salt	565 °C	Two-tank	105 MWh

Figure 2.2: Pictures of Gemasolar, left: central receiver tower with heliostat mirrors in standby; right: *(1)* hot tank, *(2)* cold tank, *(3)* heat exchanger, *(4)* power block [21]

all the solar tower systems are planned with a storage unit. Currently, only 401 MW from central receiver CSP are currently installed without any storage. Among those, Ivanpah with 392 MW has the largest share, as its special location close to Las Vegas makes it possible to use the midday sun directly for air conditioning. Only 121 MW are in construction (Ashalim in Israel) without a storage unit [22]. All of those plants without storage use water/steam as the heat transfer fluid; therefore, energy storage, e.g. in pressured vessels, is not that easy and cheap.

All planned solar tower plants will include a storage unit. Furthermore, it is interesting to note that currently, different heat transfer fluids (molten salt, water/steam, liquid sodium and air) and therefore different storage systems are in operation (Figure 2.3). However, all the planned storage central receiver systems – with the exception of Jemalong 2 – will use molten salt as the heat transfer and storage medium in two-tank arrangements. The installed (electrical) power will increase to a higher range, from 50 MW up to 450 MW. In total, 224.5 MW are currently installed in solar tower systems having a storage unit, 1105 MW are under construction and 1612 MW are under development.

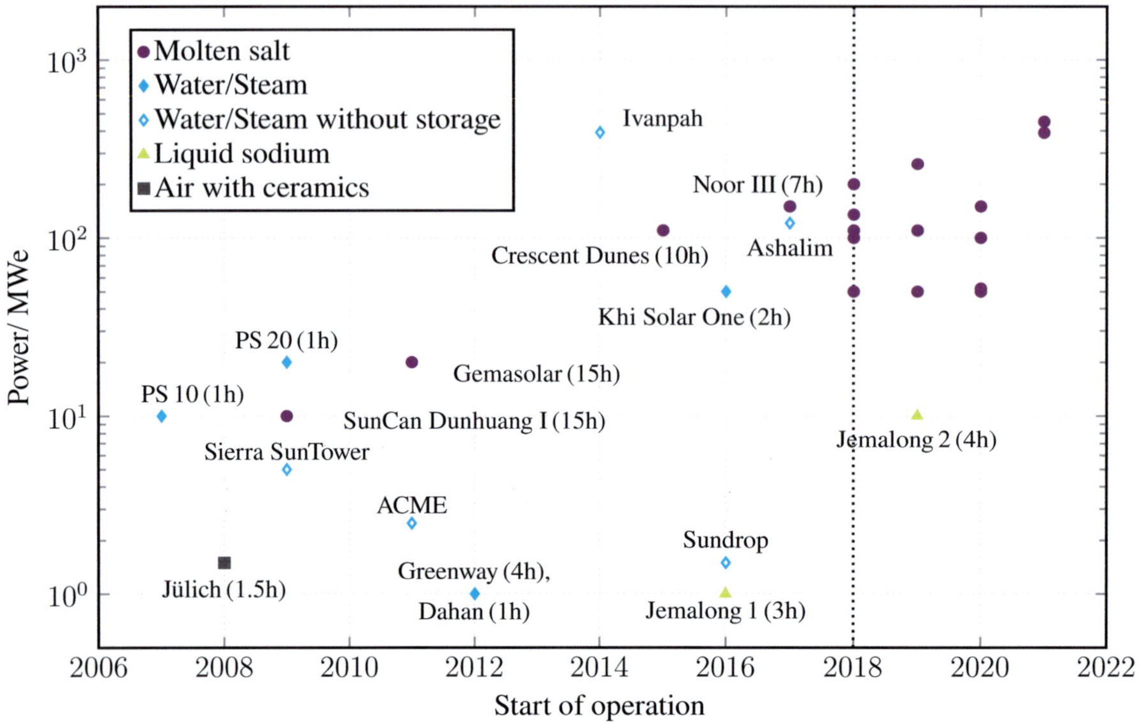

Figure 2.3: Installed electric power in solar tower plants up to 2018 [22, 23]

Figure 2.4 shows the planned central receiver plants with an electric output of 50 MW and higher. Therefore, Jemalong 2, with liquid sodium as the heat transfer medium, is not included in this graph. The solar tower plants with the highest electrical output are planned in Chile (Copiapo, Tamarugal and Likana) with storage times of 13–14 h. Furthermore, Atacama-1 and Atacama-2 are planned with high storage times of 17.5 h and 15 h. China plans several projects in the range of 50–200 MW, Golmud being the largest with 200 MW. In Marocco (Noor III), Australia (Aurora) and the United Arab Emirates (DEWA) solar power plants are in development in the range of 150–200 MW.

In Figure 2.5 the installed (*blue*) and planned (*red*) solar tower plants with storage units in MW (electric output) are summarized for each country.

The largest amount of solar electricity is planned in Chile (1320 MW). In China, 845 MW are under construction and development. In the USA, by contrast, no further plants are currently planned. An overview of the installed and planned solar tower systems can be found in the Appendix (Section A.1).

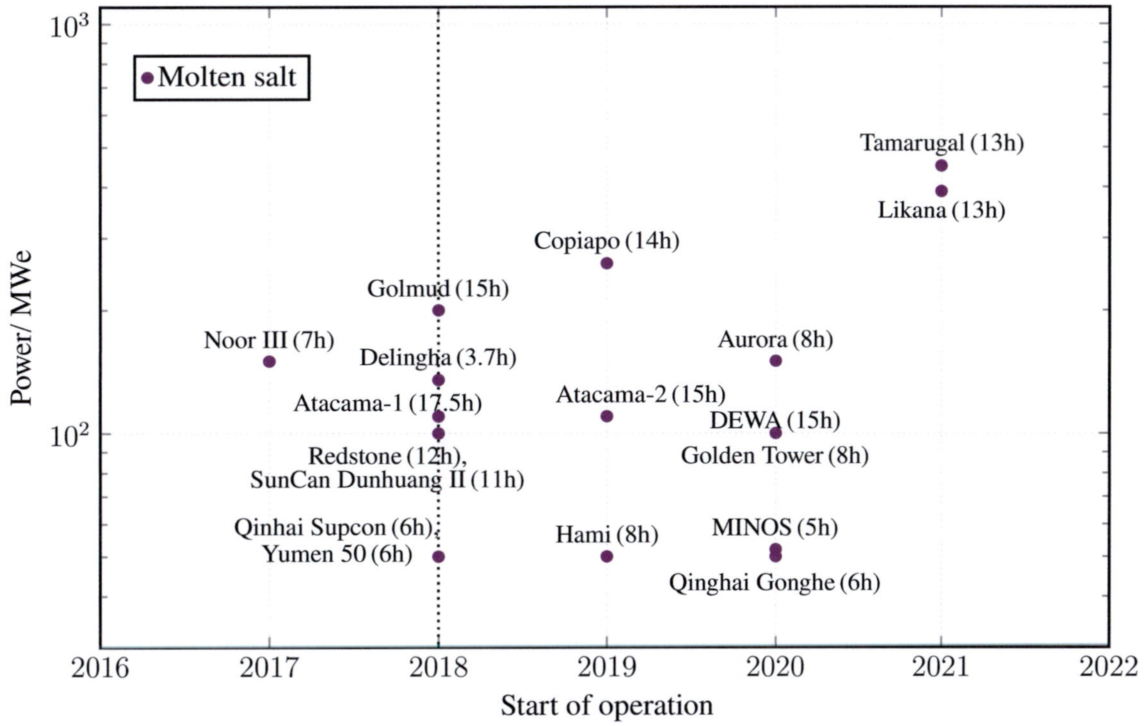

Figure 2.4: Planned electric power in solar tower plants with molten salt as heat transfer fluid and a storage system [22, 23]

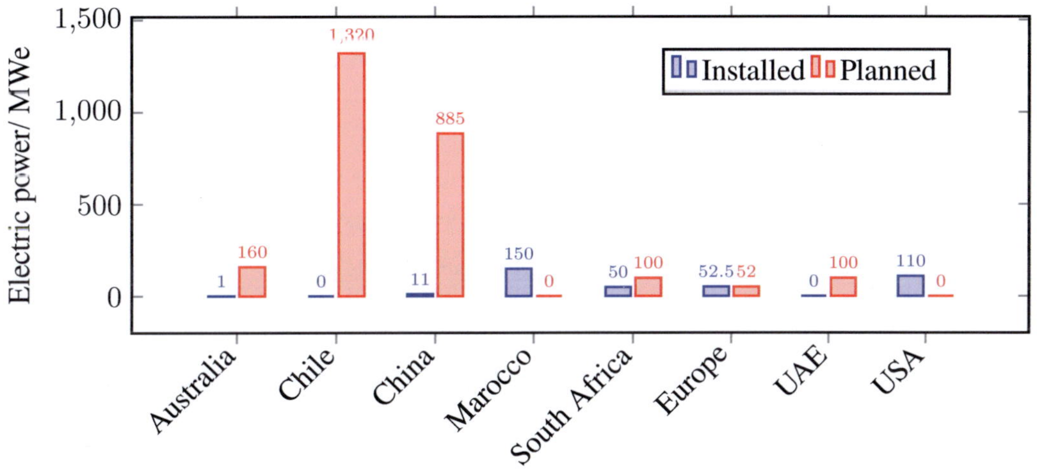

Figure 2.5: Installed and planned electrical power from central receiver CSP including a storage system for each country [22, 23]

2.1.2 Competition with photovoltaics

Solar electricity can also be produced by using photovoltaic (PV) technology. At the end of 2016, global PV capacity was 300 GW as compared with 5 GW CSP capacity [1]. Currently, electricity from PV is less expensive than that from CSP, as shown in a study by Fraunhofer ISE (2013) (Figure 2.6). However, CSP can be competitive with PV if the storage unit is included in the comparison, due to the significantly higher investment costs for battery- compared to thermal-storage systems.

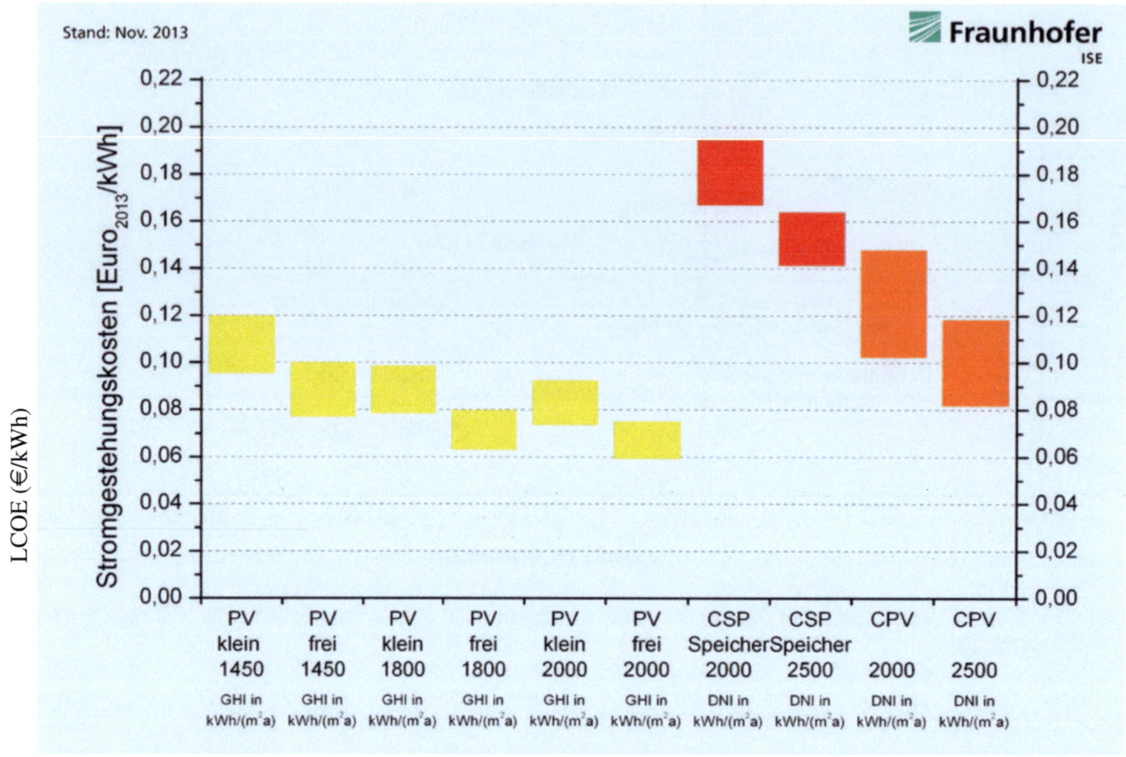

Figure 2.6: Comparison of LCOE of PV and CSP electricity depending on the solar irradiation (2013) [24]

Electricity storage world-wide is covered mainly by pumped-storage hydroelectricity; only a small percentage is stored using other technologies, such as compressed air energy storage, sodium sulphur and lithium-ion batteries, all of which are still at the demonstration level [15, 25]. Schmidt et al. [26] project an estimated price range of USD340 $\pm$ 60/kWh for stationary systems with capacities of 1 TWh .

Table 2.2 lists the projected levelized cost of electricity (LCOE) for a PV plant including storage and a CSP plant including storage in 2020 reported in recent studies. The LCOE is defined in Equation 2.1.

$$\mathrm{LCOE} = \frac{I_0 + \sum_{t=1}^{n} \frac{A_\mathrm{t}}{(1+r)^t}}{\sum_{t=1}^{n} \frac{E_\mathrm{t}}{(1+r)^t}} \tag{2.1}$$

It is determined by dividing the sum of the investment costs I_0 and the total costs A_t (including operationing and maintenance costs) by the electricity E_t generated in a year t. The factor r stands for the interest rate in a year t. The influencing factors are summed up for the life span of the plant n [24].

Table 2.2: Comparison of LCOE of PV+storage with CSP+storage from different references, *10yr, +15yr

Ref.	Scale (h)	LCOE PV+st. (US$/MWh)	LCOE CSP+st. (US$/MWh)	in year
Finkel [27]	3 (PV), 12 (CSP)	105	130	2020
Lazard [28]	10	92	182	2016
NREL [29]	9	170*/150+	110	2020
NREL [29]	3	140*/120+	120	2020

The figures in Table 2.2 show a large discrepency. For large-scale storage (9–12 h), CSP can be either cheaper (≈ -40%) than PV+storage in 2020 according to the NREL study [29] or significantly more expensive according to the Lazard review [28] (≈ +100%) and approximately in the same range according to the Finkel review [27] (≈ +20%). For a smaller thermal energy storage system of 3h, the NREL reports quite similar results for the costs of PV+storage and CSP+storage [29].

A different approach is to compare the "value" of PV and CSP+storage. It includes the degree of flexibility regarding dispatchability to the grid [4]. The total value is a combination of the operational and capacity value. Operational value stands for the costs that are avoided during the operation of the grid (fuel, start-up, maintenance, emission). Capacity value represents the costs that can be saved by avoiding the need to build new conventional generators due to an increased energy demand [30]. A study of NREL compares CSP+storage with PV regarding the operation and capacity with different percentages of renewables in California. Figure 2.7 shows that the gain in value of the electricity from CSP including a storage unit compared with PV can be ≈ US$ 60/MWh in the presented case [30].

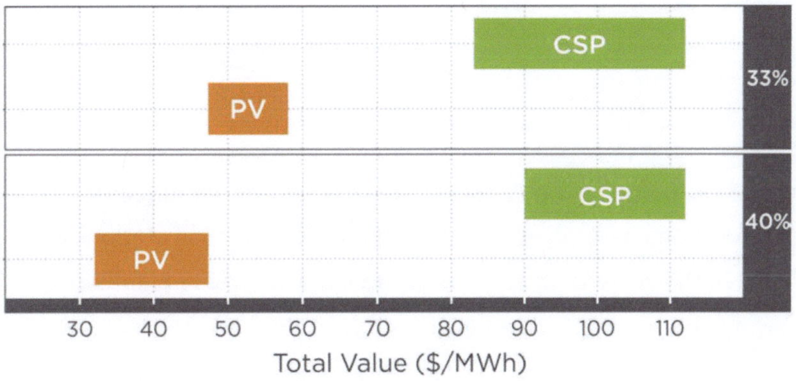

Figure 2.7: Value of PV vs. CSP+storage with 33% and 40% of renewables in California [4]

2.1.3 Cost-reduction goals for CSP

From the projections described, it can be concluded that the LCOE for CSP with storage needs to be significantly reduced. The US Department of Energy (DOE) sets the LCOE goal for 2020 at US¢ 6/kWh including thermal storage with US¢ 1/kWh in the Sunshot Vision Study of 2012 [3]. This includes the targeted investment costs of the storage system of US$ 15/kWh. The International Energy Agency (IEA) projects a higher average LCOE for 2020 for a CSP plant with storage of US¢ 13/kWh. For 2050, US¢ 7/kWh are envisaged in the Technology Roadmap Solar Thermal Electricity of 2014 [17]. One key parameter in achieving this goal is to reduce the investment costs of the storage system to 15 €/kWh according to the Strategic Research Agenda of ESTELA [31], which is in the same range as the DOE target.

Lilliestam et al. [32] show that the LCOE for solar towers was in the range of US¢ 17–26/kWh (2007–2015) and are now reduced to US¢ 6–16/kWh (2015–2017), which already comes close to the DOE value. Table 2.3 presents an overview of a few selected solar tower plants including a thermal energy storage system that have most recently been put into operation or are planned to be operating within the next 5 years.

Table 2.3: Power purchase agreement prices of operational and planned CSP plants including a thermal energy storage unit [4]

Project	Country	MWe	Storage Capacity (hours)	Power Purchase Agreement Price (PPA) (¢/kWh)	Status	Completion Date
Crescent Dunes	U.S.	110	10	13.7	Operation	Q4 2015
Noor III	Morocco	150	7.5	16.3	Construction	Q4 2017
Redstone	South Africa	100	12	12.5 (PPA to be signed)	Development	Q3 2018
DEWA CSP Project Phase I	United Arab Emirates	200	12	8.0 (targeted)	Planning	Q2 2021
Copiapo	Chile	240	14	6.3 (bid)	Planning	TBA

The power purchase agreement price of these plants is within the range of ¢ 6.3–16.3/kWh with a decreasing price for the newest plants. In the IRENA Report [33] on power generation costs in 2017, similar trends in the auction prices are reported, down to US¢ 8/kWh in 2022.

2.2 Recent developments in thermal energy storage

This section gives a general overview of the three main types (sensible, latent and thermochemical) of thermal energy storage and the related scientific work in the recent past[1]. The research activities are oriented towards reducing the costs per unit stored energy compared with those of the commercially applied two-tank configuration, in order to improve the competitiveness of the CSP industry (Section 2.1).

2.2.1 Sensible thermal energy storage

Sensible thermal energy storage systems use a temperature difference between the charged and the discharged state (Equation 2.2). The density ρ and the specific heat capacity c_p of the stored medium determine the volume needed for storing a certain amount of energy Q_sens, assuming a fixed temperature difference ΔT.

$$Q_\mathrm{sens} = c_p \cdot \rho \cdot V \cdot \Delta T \tag{2.2}$$

The thermal energy can be stored either directly or indirectly. In a direct system, the heat transfer fluid acts also as the storage medium. In the indirect case, the heat transfer fluid transfers heat to a storage material, e.g. using a heat exchanger. Indirect storage has the disadvantage that exergy is lost due to the temperature gradient needed to transfer the heat to the storage medium during the charging process and back during the discharging process. However, if the storage material is significantly cheaper and has a much higher storage density than the primary fluid, it can still be advantageous. Figure 2.8 shows the schemes for direct and indirect storage.

Figure 2.8: a) (Two-tank) direct system; b) (two-tank) indirect system

In sensible thermal energy storage systems, cost reductions can be achieved either by reducing the costs of the material (tank/ storage material) or by increasing the stored energy per volume, i.e. reducing the amount of storage material while maintaining the same energy. The approaches addressing these issues are shown in the following.

[1] The results of this section are adapted from a literature research that has been conducted within the framework of the Bachelor thesis of M. Pfuhl (2015).

Reduce two tanks to one tank – avoid dead volume. The idea is to reduce the amount of tank material and avoid unused volume in the tank. Two tanks with cold and hot material are combined to one single tank (Figure 2.9).

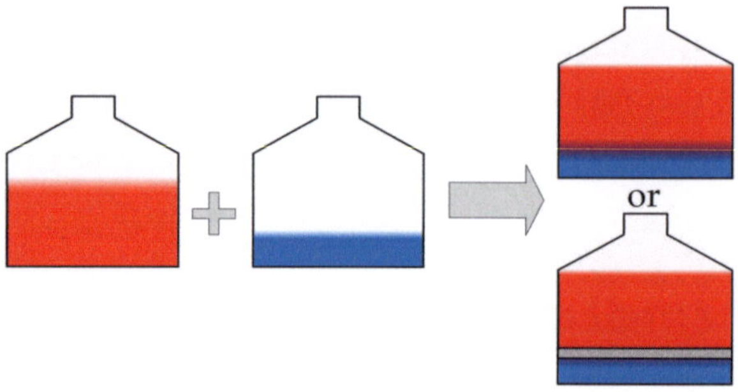

Figure 2.9: Principal scheme of reducing two tanks to a single tank either with or without a mechanical barrier to separate the cold and the hot part

The hot and cold parts are separated by their density difference or by a moveable mechanical barrier. Now, no empty volume is present, in contrast to the two-tank storage, where, for example, at the fully charged state the cold tank is almost empty and the hot tank is almost completely filled. The tank material makes up approximately 25% of the investment costs according to Ref. [34] or 16% according to Ref. [35]. Querol et al. [36] report additional advantages of this storage arrangement: thermal losses are reduced and the extended shaft pumps are no longer necessary. Systems without a moving barrier have been tested in the past [20]; experiments with a moving barrier are planned but no results have yet been reported [36, 37].

Reduce costs of storage material and increase storage density by using solid filler material. Even further cost reductions are possible by replacing parts of the fluid with a solid filler material (Figure 2.10). In the best case, this filler material is cheap and has a high volumetric thermal capacity (ρc_p), so that the tank size can be reduced as well. The feasibility of such a system is demonstrated with salt/rocks/sand and oil/rocks/sand [38, 39]. Pacheco et al. [39] show that the costs can be reduced by a third compared with the two-tank arrangement, owing to reduced costs of tank and storage material.

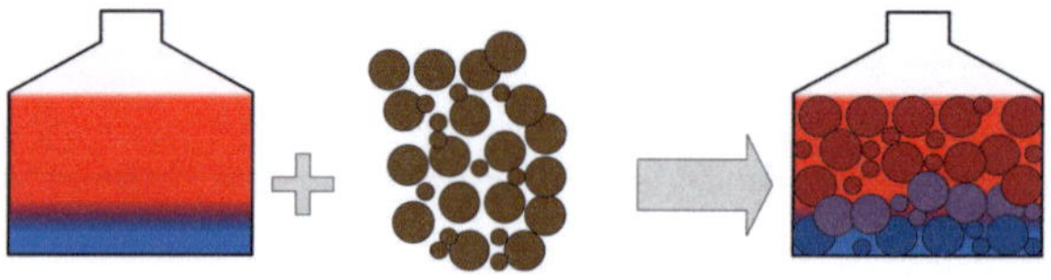

Figure 2.10: Principle scheme of reducing two tanks to a single tank with and without a mechanical barrier to separate the cold and the hot parts

There are also other approaches to using solid material as storage medium which are not discussed in detail in this work, as the heat transfer fluid is not a liquid medium and therefore not applicable to liquid metal: On the one hand, fluidized systems [40] or moving-bed heat exchangers [41] are being investigated which use particles both as storage and heat transfer media. On the other hand, packed-bed storage systems and other solid regenerator-type storage systems are under development that use a gas as the heat transfer medium [42].

Raise temperature level of storage fluid. The reason for an increase in the upper temperature limit, which is currently $\approx$ 565 °C for solar salt in solar tower plants, is based on the goal of improving the efficiency and reducing the costs of the overall CSP system. Therefore, novel salt compositions and alternative fluids are being investigated. In the Concentrating Solar Power Gen3 Demonstration Roadmap [4] salt compositions are proposed for application in the temperature range of 500–700 °C, with the heat being used for an advanced power cycle. In a report from NREL, Jonemann et al. [43] tested a chloride molten salt with a temperature range of 300–700 °C. Chloride salts, however, have the disadvantage of being highly corrosive. Fritsch et al. [44] present liquid metals as heat transfer fluids for higher temperatures than are possible with solar salt.

2.2.2 Latent thermal energy storage

In latent thermal energy storage systems, energy is stored and released using the melting/crystallization enthalpy during the matérial s phase change (mainly solid-liquid [45]). The density ρ and the enthalpy Δh_{melt} define the volume necessary for storing a certain amount of thermal energy.

$$Q_{\mathrm{lat}} = \rho \cdot V \cdot \Delta h_{\mathrm{melt,m}} \tag{2.3}$$

The energy can be stored either in an active system - the storage medium is moving - or in a passive system - the storage medium is stationary (Figure 2.11).

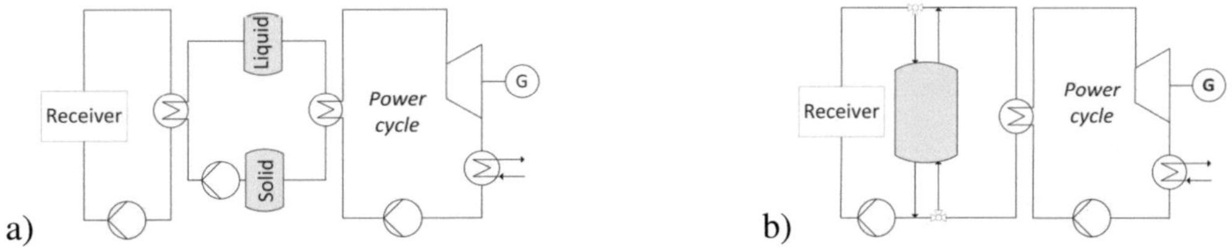

Figure 2.11: a) active latent system; b) passive latent system

The latent storage systems already have a higher storage density than most sensible systems, which can lead to cost reductions per energy unit. The main challenge for the latent thermal energy storage systems is to test candidate materials and improve the heat transfer between storage material and heat transfer fluid.

Test possible filler materials and reversibility. In the literature, various phase-change materials (PCMs) are proposed and tested [46, 47, 48, 49]. Figure 2.12 gives a schematic overview of PCMs and their melting enthalpies and temperatures. Poor stability of the latent storage material properties during cycling are reported, as are corrosion issues between the storage material and the tank. Kotzé [50] proposes metals as PCMs, as they have a higher thermal conductivity and a smaller volume change during phase change compared with salt, both of which could improve the cycling.

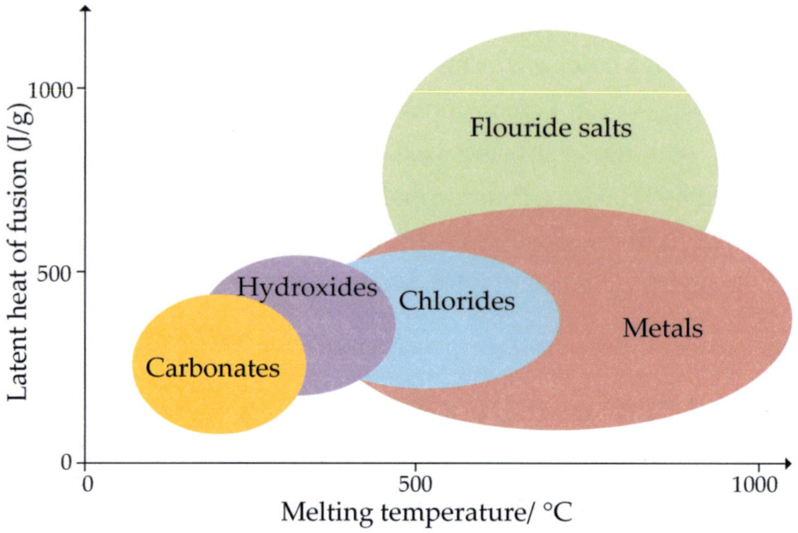

Figure 2.12: Schematic overview of PCMs according to their latent heat of fusion and melting temperature (adapted from Ref. [11])

Improve heat transfer by increasing the heat transfer surface. During discharge, the heat is extracted from the PCM, which therefore solidifies. Thus, the heat transfer is hindered at the heat transfer surface by this forming solid phase. Figure 2.13 presents three approaches to addressing this issue: finned tubes, encapsulation and active systems. These approaches are explained in detail in Ref. [35].

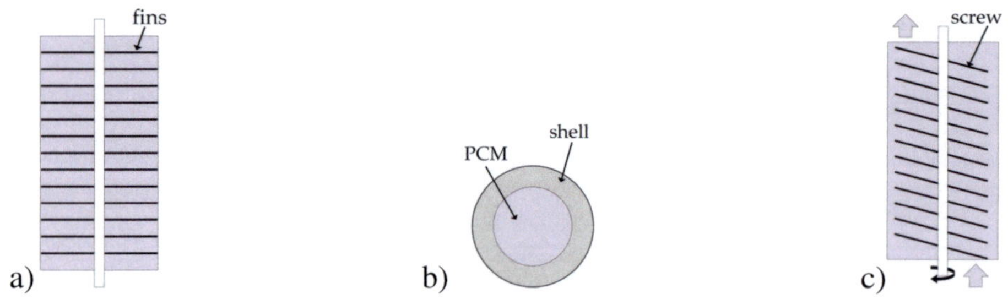

Figure 2.13: Principle schemes of latent thermal energy storage systems with increased heat transfer surface, a) finned tubes, b) encapsulation, c) active system

A finned tubes solution (Figure 2.13a) was investigated by Laing et al. [51]. Radial and longitudinal aluminium fins are proposed as the most cost-effective design. A pilot test has been successfully performed with a thermal capacity of 700 kWh using sodium nitrate salt (T_{melt} = 305 °C). Johnson et al. [52] developed a special fin design for latent storage for application in a cogeneration plant. The second approach is the encapsulation of PCM (Figure 2.13b). The heat transfer surface is significantly increased and no heat exchanger is needed. However, the heat tranfer is now limited by the thickness and the conductive properties of the encapsulation material [53]. Therefore, encapsulation and latent material are developed and tested. For example, Gimenez-Gavarrel and Fereres [54] develop glass capsules and test them with both nitrate salt and lead as the PCM. Alam et al. [55] use an inert inner polymer coating in a metallic capsule. This latent "filler" material can also be combined with solid filler material in a packed bed, as theoretically shown by Galione et al. [56]. The third idea described here is the use of active systems, meaning the PCM is transported along the heat transfer surface (Figure 2.13c). Zipf et al. [57] present theoretical investigations of the cost and performance of a so-called screw heat exchanger, where the storage material is moved by rotating screw shafts. A further active concept has been developed by Pointner et al. [58, 59], where the storage material is transported in containers along the heat-exchanging surface.

Improve heat transfer by increasing the thermal conductivity. Another approach is to increase the thermal conductivity of the latent storage material by adding highly conducting materials. For example, Mettawee et al. [60] use aluminium as an additive and show a significantly shorter charging process. Zhao et al. [61] and Mehling [62] add graphite to the PCM and show improvements of the thermal conductivity and therefore of the heat transfer performance.

Adjust temperature misalignment between latent and sensible. As latent storage systems exchange heat at constant temperature during melting/crystallization (T_{melt}), they are best suited for direct steam generation solar thermal plants. However, if the heat transfer fluid experiences no phase change, the temperature profiles are misaligned. To adjust this, cascaded latent storage systems are being investigated, where PCMs are combined and arranged according to their melting temperature, so that they fit better to the sensible temperature profile and less exergy is lost (Figure 2.14).

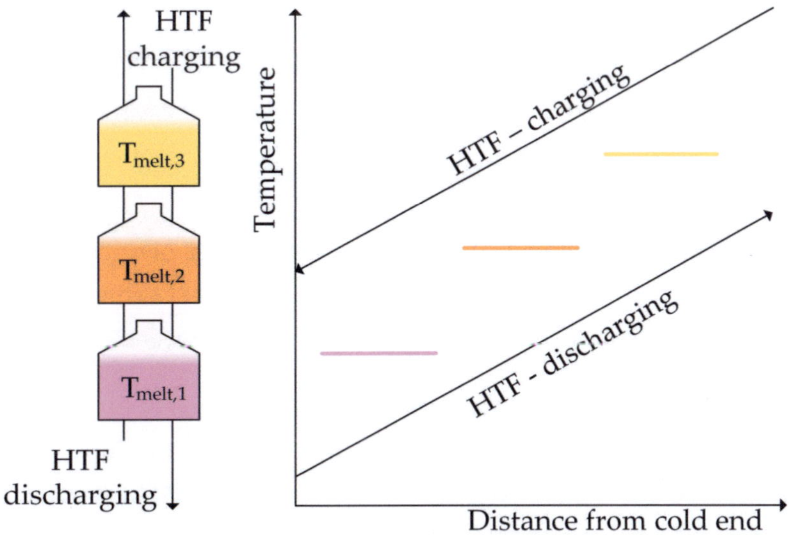

Figure 2.14: Principal scheme of a cascaded latent energy storage system and the corresponding temperature profiles with $T_{melt,1}<T_{melt,2}<T_{melt,3}$ (adapted from Ref. [63])

For example, Michels and Pitz-Paal [63] report a higher efficiency and a more uniform outlet temperature with three latent materials in cascade for the application in a parabolic through plant compared to a non-cascaded latent storage based on numerical and experimental results. In a numerical study by Elfeky et al. [64], a higher overall efficiency of a cascaded arrangement of three PCMs compared with only one is reported.

2.2.3 Thermochemical thermal energy storage

In thermochemical storage systems, the reaction heat of an endothermic process is used to charge the storage, whereas the heat released during the exothermic back reaction can be used in the discharge process. The density ρ and the reaction enthalpy $h_{\mathrm{reac,m}}$ determine the volume of the storage material needed for the amount of thermal energy stored.

$$Q_{\mathrm{thch}} = \rho \cdot V \cdot \Delta h_{\mathrm{reac,m}} \tag{2.4}$$

The principal scheme is shown in Figure 2.15. In the first reactor, heat is transferred from the receiver into the reactor to induce the endothermic reaction. The products can then be stored and used later on in the second reactor, where the heat is released to be used in the power cycle.

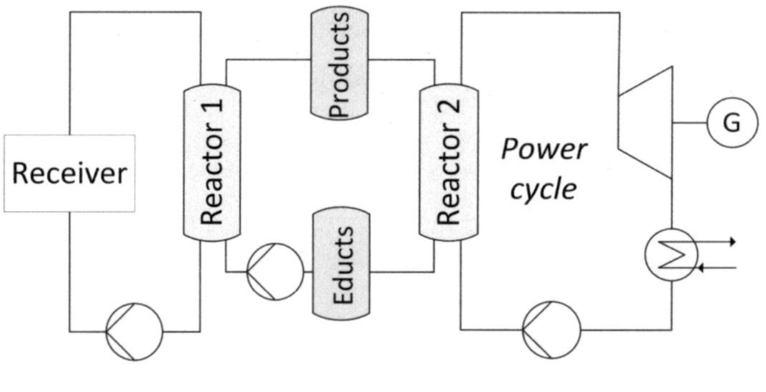

Figure 2.15: Thermochemical system

The thermochemical storage systems have an even higher storage density than the latent storage systems; however, they are still at a very low level of technological readiness. Therefore, the main research topics concern testing suitable reaction partners and the reversibility on lab scale.

Test candidates and reversibility. In the review articles of Pardo et al. [65] and Wu and Long [66] possible reactions are listed and evaluated that could be utilized as thermochemical storage in a concentrating solar thermal system. According to Pardo et al. [65], the following reactions are particularly interesting for high-temperature heat storage owing to their comparably high level of technological readiness and low cost:

- Dehydrogenation of metal hydrides: $MgH_2 \rightleftarrows Mg + H_2$
- Ammonia dissociation: $2NH_3 \rightleftarrows N_2 + 3H_2$
- Dehydration of metal hydroxides: $Ca(OH)_2 \rightleftarrows CaO + H_2O$
- Decarboxylation of metal cabonate: $PbCO_3 \rightleftarrows PbO + CO_2$

2.3 Potential use in other industries

Thermal energy storage systems are also interesting for energy-intensive industries, e.g. the steel or cement industry, where heat is generated at high temperature and not yet used owing to characteristics or requirements of the processes, so-called waste heat or excess heat [67].

2.3.1 Conditions for re-use of excess heat

Brueckner et al. [68] distinguish between theoretical, technical and economic potential of integrating heat recovery in a process, meaning that integration may be theroretically possible, but not technically feasible. Or it is theoretically and technically feasible, but not interesting from an economic point of view. All three conditions must be fulfilled to integrate a heat recovery system. Waste heat can be re-used for heating or for mechanical or electrical work, e.g. pre-heating combustion gases, feedwater furnace loads and generation of electricity [69]. According to a report from Fraunhofer ISI [70], the following factors have to be considered for the assessment of the potential to re-use waste heat in industrial processes:

- Temperature level and difference between heat source and sink
- Quantity of thermal energy
- Chemical composition (due to corrosive components above saturation temperature)
- Type of heat loss (radiative, convective)
- Local proximity of source and sink
- Synchronized source and sink

2.3.2 Thermal energy storage for re-use of excess heat

Fernandez et al. [71] name five different technologies to re-use excess heat: Heat exchangers, sorption systems, mechanical vapour compressors, power cycles (e.g. steam Rankine) and thermal energy storage technologies. Only thermal energy storage will be discussed here in detail. It allows decoupling of the heat source and sink both in time and in location.

In the reports of Fraunhofer ISI [70] and of the U.S. Department of Energy [69] the following industries with a high relevance for waste heat recovery at elevated temperatures are listed: iron and steel, metal (non-iron), cement, glass, and chemical. They are presented in the Appendix (Section A.2). In this section, examples are presented where a thermal energy storage system is already implemented.

Fernandez et al. [71] highlight the use of thermal energy storage in three industrial branches: the steel industry, waste incineration/PVC drying and glass production. Concerning the steel industry, the transportation of waste heat for re-use in district heating is discussed. Sensible (via underground piping), latent and thermochemical storage (via transportation by train) systems are compared regarding their costs, the thermochemical system being the cheapest at < 10 €/MWh owing to its large storage density. Secondly, a case is presented where waste heat from an incineration plant in Hamm, Germany, is stored thermochemically in a zeolite sorption storage, transported and re-used in a PVC drying facility 7 km away. It is charged by hot dry air and discharged with moist air, which is thereby dried and heated up. The third case shows a regeneration system in a glass furnace installed in 1857. Refractory material acts as a sensible storage, which is charged by hot exhaust gases and discharged by pre-heating the combustion air.

A further study was conducted by DLR and Badische Stahlwerke, where the addition of a molten salt storage to an electric arc furnace (steel industry) to increase both the efficiency and the electricity production was investigated [72]. In this process, 20% of the produced heat is lost in exhaust gases of the oven. This heat could be stored, transferred to a steam power block and re-used to generate electricity. A two-tank storage system similar to the one already installed in operational concentrating solar power systems is considered (T = 225–400 °C) Simulations of a storage system with a thermal power of 10 MW and a capacity of 5.3 MWh have already been performed.

In a cogeneration plant of the company STEAG in Wellesweiler, Germany, a latent storage system using salt (T_m = 306 °C) was developed in collaboration with DLR [73]. If the gas turbine fails, the storage system is able to generate steam for 15 min until the secondary boiler is ready. This reduces the amount of fossil fuels needed, as the boiler does not have to be on standby.

2.4 Conclusion

Thermal energy storage plays a crucial role in CSP technology, as CSP can be competitive with PV only if the technologies are compared including a storage unit. The presented studies show that, among others, the costs of the storage unit have to be reduced significantly to meet the targets for the overall price of CSP electricity. However, the commercially used two-tank option with solar salt does not allow for such extensive cost reduction. Thus, alternative solutions are currently being investigated, such as sensible systems with filler material, latent and thermochemical storage systems. Additionally, the potential of including a high-temperature thermal energy storage in certain energy intensive industries is shown. Especially for temperatures beyond 600 °C, where molten salts are no longer applicable, liquid metals are promising candidates as heat transfer and storage fluids.

In the following section, the presented thermal energy storage systems are considered with regard to the coupling with sodium, and the related benefits and limits are discussed.

3 Selection of a suitable storage system with sodium

As discussed in Section 1, the goal of this work is to select a promising storage system for a CSP plant with sodium as the heat transfer fluid and to determine its performance. In this section, a top-level evaluation is performed, in which a wide range of storage options are considered and narrowed down to the most promising ones.

The evaluation is conducted based on five criteria, as shown in Section 3.1. The storage options are taken from a literature search of state-of-the-art storage system that are already working in operational CSP plants, as well as storage systems which have so far been tested only on lab- to pilot-scale (Section 3.2). The results of the evaluation are described in Section 3.3. The evaluation procedure and the results are presented in detail by Niedermeier et al. [74]. They are updated and summarized in this section.

3.1 Methodology

Five criteria were selected to evaluate the thermal energy storage systems, in accordance with Refs. [20, 35]:

- Storage medium cost (investment cost for the raw material)
- Storage density (energy per volume or mass)
- Cycling behaviour (reversibility of charging and discharging cycles)
- Technology readiness level (state-of-the-art, pilot scale, lab scale)
- Suitability for sodium (compatibility issues, risk estimation)

Where possible, quantitative comparisons are conducted; otherwise, a qualitative assessment was made.

3.2 Evaluated systems

Three types of thermal energy storage systems are evaluated here: sensible, latent and thermochemical storage systems for working temperatures of 600 °C and above. For each storage system, sodium is the heat transfer fluid in the primary loop. The principle of each system and its technical design are explained in the following.

3.2.1 Sensible storage systems

The following sensible thermal energy storage systems are evaluated:

a) Direct two-tank sodium system
b) Direct single-tank sodium system with filler
c) Direct single-tank sodium system with a floating barrier
d) Indirect two-tank salt system
e) Indirect single-tank salt system
f) Indirect with gas as secondary heat transfer fluid and solid storage

The principal schemes of each arrangement are illustrated in Figure 3.1. The direct two-tank system is installed in currently operational solar tower plants, e.g. Crescent Dunes (10 hours) or Gemasolar (15 hours). To reduce tank material costs, single-tank systems are proposed in the literature [39]. By adding solid filler material, the storage medium costs can be reduced and the storage density can be improved. Indirect two-tank storage systems are also already commercially used in parabolic trough plants, where thermal oils are used as the heat transfer fluid and the heat is transferred to molten salt as the storage medium. The possibility of exchanging heat to a secondary heat transfer fluid, which transfers the heat to a solid storage, is also included in the study. For detailed information on the systems, refer to Niedermeier et al. [74].

3.2.2 Latent storage systems

Three latent thermal energy storage systems are evalated (Figure 3.2):

a) Finned-tubes system
b) Packed-bed system of encapsulated PCM
c) Active system: e.g. screw type

A drawback to the latent-heat storage systems is that, during discharge, the growing solid layer on the heat transfer surface builds a thermal resistance. Therefore, all the presented latent storage systems are supposed to either maximize the heat transfer surface or to move the latent storage material along the heat transfer surface.

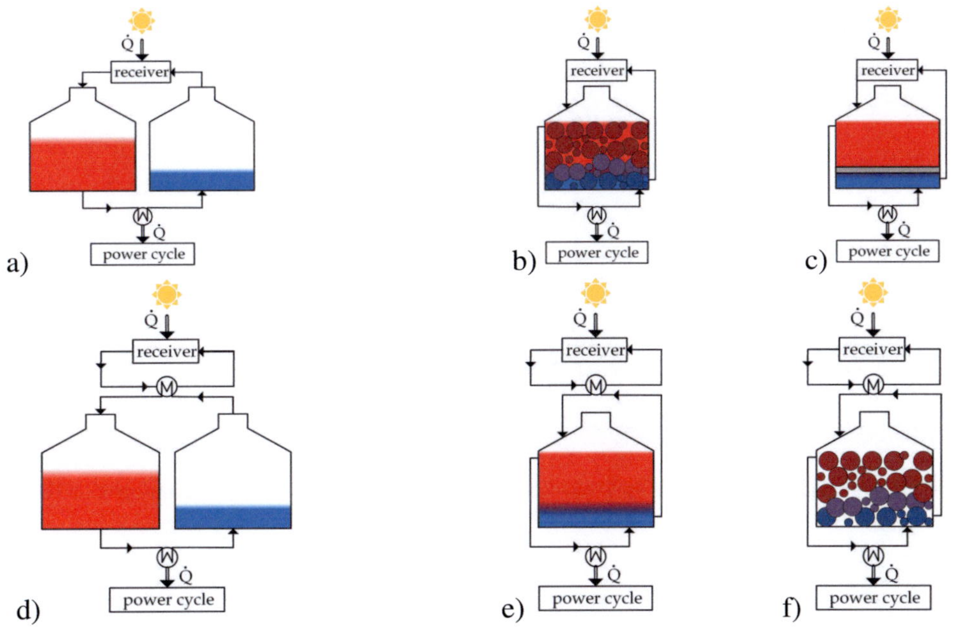

Figure 3.1: Evaluated sensible systems: a) direct two-tank, b) direct single-tank with filler, c) direkt single-tank with a floating barrier, d) indirect two-tank, e) indirect single-tank, f) indirect with gas as secondary heat transfer fluid

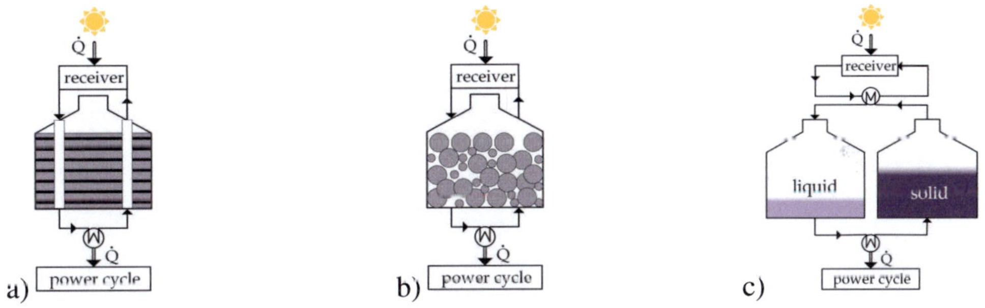

Figure 3.2: Evaluated latent systems: a) finned-tubes, b) packed-bed of encapsulated PCM, c) active: e.g. screw type

3.2.3 Thermochemical storage systems

Four thermochemical thermal energy storage systems are evaluated (Figure 3.3):

a) Dehydrogenation of metal hydrides
b) Ammonia dissociation
c) Dehydration of calcium hydroxide
d) Decarboxylation of lead carbonate

These four reactions are selected because of their reaction temperature and because their use for thermal energy storage for CSP plants has been reported in the literature [65]. The metal hydrides release hydrogen with heat input. Hydrogen can be stored as gas or transferred to another metal until the heat is again given to the power block and the hydrogen is transferred back. The dissociation of ammonia (with heat input) to nitrogen and hydrogen and the synthesis (under heat release) is well-known as the Haber-Bosch process. Calcium hydroxide reacts to water and calcium oxide in the endothermal reaction, heat is released in the reversed reaction. Lead carbonate reacts to lead oxide and carbon dioxide with heat input.

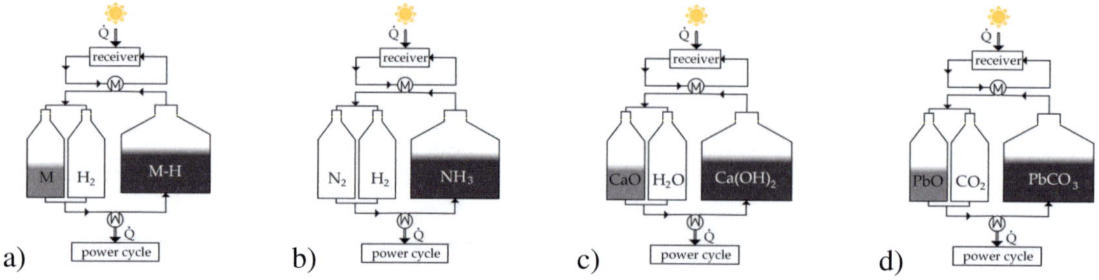

Figure 3.3: Evaluated thermochemical systems: a) metal hydrides, b) ammonia, c) calcium hydroxide, d) lead carbonate

3.3 Results

The criteria listed previously in Section 3.1 are now applied to the proposed storage systems. The systems are compared with the state-of-the-art two-tank system with solar salt. For a more detailed evaluation, see Niedermeier et al. [74].

3.3.1 Sensible storage systems

Sensible thermal energy storage in general is the most mature technology compared with latent and thermochemical. It is already used in currently operational power plants as a two-tank configuration. Sensible storage systems are simple systems and the storage medium costs per unit mass are relatively low. However, the storage density is also relatively low, which leaves few possibilities for cost reductions.

The direct system with sodium can be realized in a two-tank and a single-tank arrangement. In both cases, there is the potential risk of a reaction with steam if there is a leakage in the heat exchanger. The two-tank system is the simplest one, as the cold and hot fluid are separated in different tanks. Thus, a uniform fluid outlet temperature can be realized during discharge. The cycling (charge/discharge) is well-developed in operational CSP plants with solar salt. Moreover, there is already some operational experience with such a configuration with sodium within the frame of the experiments in the IEA-SSPS in Almeria. However, there is limited potential for further cost reduction. And, as there is a large amount of sodium in the system, there is a high risk of potential damage.

The single-tank configuration has the advantage of replacing two tanks, which are partially empty the whole time, by a single one, which is completely filled. The hot and cold parts of the fluid are separated either by a temperature gradient only, leading to a density gradient in the tank separating cold and hot part, or by a mechanical barrier (floating barrier). Further cost reduction can be reached by adding low-cost solid filler material and improving the storage density at the same time. A system with a porosity of 22% can be 72% cheaper in terms of €/kWh (compared with a direct two-tank system with sodium) and 11% cheaper compared with an indirect two-tank molten salt system, if only storage medium costs are considered. With temperatures higher than 565 °C, indirect storage with solar salt is no longer feasible. There is already practical experience on a pilot scale with molten salts [39] and thermal oil [38] as fluids in packed-bed systems. By adding filler material the amount of sodium can be reduced significantly in the system and therefore, the potential damage in case of a sodium fire can be minimized. However, a chemically compatible filler material is required, and charging and discharging is more complex than in a two-tank configuration, as there is heat transfer between the fluid and the solid filler material, which leads to decreased efficiency compared with the two-tank arrangement. Furthermore, the good thermal conductivity can lead to a reduction of the storage capacity due to the degradation of the thermocline zone between hot and cold fluid, especially during standby. In a single tank with a moving barrier, the heat transfer from the hot to the cold zone is blocked. However, as no filler material can be used, high amounts of sodium are still needed, meaning higher costs and higher potential risks compared with a single tank with filler material.

As indirect systems with molten salts, the commercially available two-tank configuration and the single-tank system are evaluated. As both are indirect systems, a heat exchanger is necessary between sodium and the storage fluid, which is also a cost factor. However, it can also be advantageous to decouple the sodium loop from the steam cycle. The two-tank system with solar salt is the one with the highest technology readiness level and lowest complexity. However, the storage density is relatively low. Furthermore, the upper temperature of solar salt is limited and salts that are stable at higher temperatures are highly corrosive and have high melting temperatures. Experiments in a single-tank configuration with molten salt and filler material have already been conducted on pilot scale. Similar to a single-tank system with filler and sodium, compatible filler material needs to be found and the storage capacity is reduced due to the region with the temperature gradient.

Finally, solid material as the main storage material is also evaluated. The heat is transferred from sodium to a gas, which then heats up the solid material. A cheap solid material can be used, as it has to be compatible only with the gas and not with the heat transfer fluids sodium or salt. However, there are disadvantages regarding the poor heat transfer from gas to solid and reverse.

3.3.2 Latent storage systems

Latent thermal energy storage systems in general have higher storage densities than sensible ones owing to the melting enthalpy. On the other hand, the storage system is more complex than the sensible storage one, as a high heat transfer surface has to be ensured due to the relatively low thermal conductivity of the phase-change material. The systems are still on a low technology readiness level and experiments to date have been mainly on laboratory scale.

The finned tubes concept is the simplest of the proposed latent storage configurations. However, it requires a complex heat exchanger with a large surface to overcome the thermal resistance in the solidifying latent storage material during discharge.

In a packed bed with encapsulated PCM the heat is transferred without a heat exchanger, as the heat transfer fluid can flow through the packed bed. On the downside, the capsules have to be flexible to endure the volume expansion at phase change and chemically compatible with the latent storage material and sodium. Possible reactions have to be considered if there is leakage in a capsule.

In screw-type configurations the heat transfer is improved by constantly removing the solidifying latent storage material from the heat transfer surface. Therefore, a smaller heat transfer surface is necessary for a given capacity compared with the previous concepts. Owing to to moving parts, however, the system is more complex and potentially more cost-intensive.

3.3.3 Thermochemical storage systems

Thermochemical thermal energy storage systems have the highest storage densities among the systems presented. In some cases, the products can even be stored at ambient temperature and therefore no insulation is necessary. The costs of such systems are difficult to estimate, as there is a lack of cost data concerning those systems in the literature, and the costs are dominated by the reactors and not by the storage material itself. Furthermore, thermochemical storage systems have the lowest technology readiness level. The systems have been tested only at laboratory scale, and more tests are required concerning the cycling.

The dehydrogenation of metal hydrides is simple regarding product separation, as one product is solid and the other one (hydrogen) is gaseous. Also, a good cycling behaviour has been reported. However, hydrogen needs to be stored under high pressure or in another metal in a more complex process. Moreover, the dissociation at high temperature requires high pressure. Corgnale et al. [75] present a screening analysis for thermochemical storage systems based on metal hydrides and conclude that $NaMgH_3$, TiH_2 and CaH_2 are particularly interesting owing to their high volumetric energy density and operating temperatures. Ward et al. [76] then propose coupling high-temperature metal hydrides (NaMg) with low-temperature metal hydrides ($NaAlH_4$ and Na_3AlH_6) based on a techno-economic analysis. Felderhoff and Bogdanović [77] propose magnesium hydrides for thermochemical storage because of low cost and good cycling stability. Paskevicius et al. [78] list metal hydride candidates for use in thermochemical storage systems. They show results of a prototype-scale experiment with MgH_2 and demonstrate reversibility in 20 cycles.

Ammonia synthesis is a well-known process (Haber-Bosch). The resulting dissociation products, nitrogen and hydrogen, can be stored at ambient temperature. Ammonia is a low-cost chemical and liquid at ambient temperature, which makes the handling easy. However, the reaction needs high operating pressures of up to 30 MPa [79], and hydrogen and nitrogen have to be stored in special high-pressure vessels. The ammonia storage system is demonstrated by Lovegrove et al. [79] in a dish solar concentrator. In a recent work, Chen et al. [80] model the heating of supercritical steam with the thermal energy from the ammonia synthesis reactor.

For the dehydration of calcium hydroxide low-cost educts can be used and the reaction can take place at relatively low pressure. A reversibility of 20 cycles has been shown in Ref. [81]; however, no complete dehydration was achieved. Schmidt et al. [82] present results of a lab-scale reactor test based on the dehydration of calcium hydroxide to calcium oxide and water. There are issues of low heat transfer and agglommeration and sintering of the products. Furthermore, water vapour is produced in the reaction; thus, there is the potential risk of a sodium reaction in case of leakage in the heat exchanger.

The decarboxylation of cerussite (PbO_3) is suggested owing to to its high volumetric energy density and easy separation of the gaseous and solid products. Furthermore, there are no by-products and no catalyst is necessary [65]. The cerussite decarboxylation is shown by Kato et al. [83] combined with a $CaCO_3$/CaO system for application in a chemical heat pump. Apart from their experiments, only few experimental work is reported [65]. Additionally, CO_2 is produced, and the toxic heavy metal lead is part of the reactions.

3.4 Conclusion

In this chapter, sensible, latent and thermochemical thermal energy storage concepts have been evaluated based on five criteria: storage medium cost, storage density, cycling behaviour, technology readiness level and suitability for sodium. General conclusions are an increase in storage density and a decrease in cycling behaviour and technology readiness level from sensible to latent to thermochemical. Concerning the storage medium cost and suitability for sodium, no general conclusions can be drawn. Regarding the individual storage concepts, four are chosen as most promising for the application to a CSP plant with sodium as the heat transfer fluid (Figure 3.4):

a) Sensible: single-tank sodium system with filler
b) Latent: packed-bed system of PCM capsules
c) Thermochemical: ammonia dissociation/synthesis
d) Thermochemical: dehydration/hydration of calcium hydroxide

The direct sensible storage system with filler material needs no heat exchanger as the heat is transferred directly to the solid material by the fluid flowing through the packed bed. The system

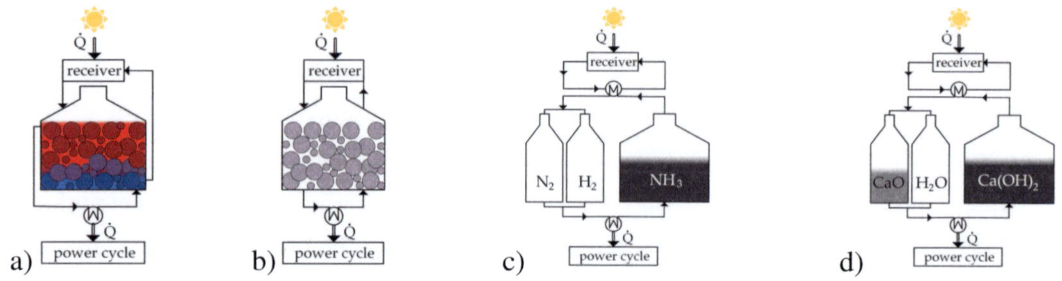

Figure 3.4: Selected promising storage systems based on evaluation: a) single-tank with filler, b) packed-bed of PCM capsules, c) ammonia, d) calcium hydroxide

has been tested successfully for molten salt and thermal oil. Furthermore, the sodium mass in the storage can be minimized by replacing parts of the storage medium with solid filler material compared with a two-tank configuration. This also results in a higher storage density (smaller tanks) and lower cost. The greatest challenge is to maintain a stratification between hot and cold material in the tank despite the high thermal conductivity of sodium. Furthermore, a suitable filler material needs to be found which is stable at elevated temperatures (600 °C and higher) and chemically compatible with sodium.

The latent storage system with a packed bed of PCM capsules is similar to the previous system, but has an increased storage density. However, at the same time, the system has a higher complexity and is at a lower technology readiness level. The development of suitable capsules is necessary. A combination of a sensible and latent packed bed is also possible and already suggested in the literature, as it combines the advantages of both systems.

The thermochemical storage system based on ammonia is promising in view of its large storage density compared with the latent and sensible systems. Although the reaction is well-known, complete reversibilty has not yet been shown. Furthermore, the high pressure required for the reaction and storage is disadvantageous.

The thermochemical storage system with calcium hydroxide is also favourable owing to its large storage density; additionally, the educts can be stored at ambient temperature. One hundred reversible cycles are shown. However, problems of agglomeration and sintering have to be solved [84].

Out of the four most promising thermal energy storage systems for a CSP plant with sodium as heat transfer fluid, the sensible packed bed system is selected in this work for further investigation. For further optimization PCM capsules could be added to the system to further increase the storage density. The suggested thermochemical storage systems should be further improved before a combination with sodium can be investigated.

In the next section, an overview of theroretical and experimental studies of the selected packed-bed thermocline storage systems in the literature is given. This will enable the detection of general gaps in this research field. Furthermore, compatible filler material for sodium are suggested.

4 Literature search regarding single-tank packed-bed storage

A search of the literature on the current status of the thermocline technology shows that mainly theoretical (Section 4.1) and only few experimental investigations (Section 4.2) of dual-media storage systems have been conducted. Most of the research articles have been published in recent years (since 2005). This shows that this research topic is up-to-date. For liquid metal neither experimental nor theoretical studies – with the exception of the publication of Pomeroy [10] – have been performed to date, to the best of the author's knowledge. Furthermore, a review of the literature concerning compatible filler material with liquid sodium is presented (Section 4.3).

4.1 Theoretical studies with conventional fluids

Four focuses of theoretical studies are presented in this section.

- Parametric studies
- Comparison of fluids
- Thermo-economic studies
- Thermo-mechanical analysis

4.1.1 Parametric studies

Yang and Garimella [85] establish a correlation of efficiency dependent on tank height and Reynolds number for molten salt as heat transfer fluid in a packed bed. The efficiency increases with increasing tank height and decreasing Reynolds number. They also show improvement of the efficiency with decreasing filler diameter in Ref. [86].

Van Lew et al. [87] investigate the influence of the tank height-to-diameter ratio, the filler particle diameter and porosity on the efficiency of a packed bed with thermal oil as heat transfer fluid. Highest efficiencies are reached with high height-to-diameter ratios, small particles and porosities between 0.15 and 0.35. Reddy et al. [88] observe the same dependence of porosity on efficiency for thermal oil, solar salt and HITEC as heat transfer fluids.

Xu et al. [89] perform an extensive parametric study of a packed bed system with molten salt as heat transfer fluid. They investigate the fluid inlet velocity, the fluid inlet temperature, the porosity and tank height, whereas the inlet velocity and fluid inlet temperature do not show a significant influence on the discharge efficiency. In a subsequent publication [90], the authors analyse the influence of the filler size and properties, with small particle diameters showing best efficiencies. They show that the filler properties are relevant only for large particle diameters.

Cascetta et al. [91] vary the particle diameter, the height-to-diameter ratio and input thermal power. The efficiency increases with decreasing diameter and increasing height-to-diameter ratio. A change of input thermal power has no significant influence on the efficiency.

All in all, the parametric studies investigate the influence of tank dimensions, porosity, filler dimensions and properties, Reynolds number, fluid inlet temperature and velocity/flow rate and input thermal power on the efficiency of molten salt with filler storage systems or thermal oil with filler. Amonge those, the tank dimensions, porosity and filler dimensions influence the efficiency significantly.

4.1.2 Comparison of fluids

Cascetta et al. [91] compare the efficiency of air, solar salt and Therminol VP-1 in a packed-bed storage, with molten salt showing highest efficiencies followed by oil and air. They define different temperature limits for each fluid, resulting in different tank dimensions for each fluid. The reference overall storage capacity is $5\,MWh_{th}$ for all heat transfer fluids. The storage efficiency is defined as the amount of total stored energy after a discharging cycle compared with the initial one.

Modi et al. [92] investigate the performance of solar salt, HITEC and Therminol 66. They keep the tank dimensions constant; thus the initial storage capacity is different for each fluid. They compare the storage capacity after discharging the storage, leading to highest capacities for solar salt, followed by HITEC and Therminol 66.

Vilella and Yesilyurt [93] perform a comparative study of solar salt, HITEC XL and Therminol in a packed-bed storage. They define the storage efficiency as energy extracted during discharge compared with that during an ideal discharge. They state that a comparison should be based on the same capacity for each fluid, resulting in similar efficiencies for all fluids. Moreover, the authors compare the storage-material costs and recommend to include the pumping losses as well when comparing heat transfer fluids.

Reddy et al. [88] compare the efficiency of Therminol, solar salt and HITEC. The efficiency is defined as by Vilella and Yesilyurt [93]. For the comparison, they keep the tank dimensions constant; thus the initial storage capacity is different for each fluid. Therminol shows highest efficiencies, followed by solar salt and HITEC.

All in all, the comparison of different fluids is not performed in a consistent way in the literature. The reference case is sometimes a tank with constant dimensions leading to different capacities for each fluid, sometimes a tank with constant capacity leading to different dimensions for each fluid. The definition of the efficiency of a single-tank storage with filler material varies, too. Furthermore, the required pumping power and the storage costs are typically not included in the comparisons.

4.1.3 Thermo-economic studies

Angelini et al. [94] carry out a performance comparison of a thermocline molten salt storage with the state-of-the-art two-tank configuration. They conclude that the performance is reduced to 64%, allowing a threshold of $T_{max} - 5$ K, compared with the two-tank configuration (100%).

Biencinto et al. [95] perform a study on the annual performance of four systems: direct and indirect two-tank storage and direct and indirect single-tank packed-bed storage. For the direct systems, Therminol VP-1, and for the indirect systems, solar salt is taken as the storage fluid. The results show that the annual electricity yield of the single tank system is always below the two-tank one. Furthermore, they reach the highest efficiencies when discharging the tank completely and extracting the whole thermocline region.

Kolb [96] shows that a similar annual performance can be reached with a single-tank configuration. However, he states that this is owing only to the flexibility of the power block of the parabolic trough plant being able to work at temperatures below design point; otherwise, the two-tank configuration would perform better.

Cocco and Serra [97] compare the performance of a single-tank thermocline storage with a two-tank system. They conclude that the two-tank configuration shows slightly higher efficiencies, but the single tank leads to lower levelized cost of electricity.

Pacheco et al. [39] determine the cost of an 688-MWh indirect single-tank system with a packed bed and compare it with an indirect two-tank system, both using molten salt as heat transfer fluid. They include the investment costs for storage medium, tank and heat exchanger. They conclude that a cost reduction of 30% can be reached with a single-tank arrangement (20 $/kWh$_{th}$) owing to reduced tank and storage material compared with the indirect two-tank configuration (31 $/kWh$_{th}$).

Strasser and Selvam [98] compare the investment cost of a 2165-MWh single-tank system with filler to a structured thermocline system with concrete and arrive at 30 $/kWh$_{th}$ and 34 $/kWh$_{th}$.

The calculation includes further parameters such as insulation, piping and engineering costs, leading to higher costs than estimated by Pacheco et al. [39].

Libby [99] calculates a reduction of the storage investment cost of 33% with a thermocline system compared with the state-of-the-art system for large capacities (3000-3500 MWh_{th}).

Gagglioli et al. [100] show a cost reduction of 39% when using a combination of single-tank storage (without filler) with steam generation compared with a two-tank arrangement (80 MWh_{th}).

In summary, the authors cited agree that cost reductions of $\approx 30\%$ are possible by using a single tank instead of two tanks and replacing large parts of the storage fluid with a filler material. The figures vary with the overall system, the storage capacity and the components included in the assessment. The efficiency of a single tank system, however, is lower and depends on the applied power block.

4.1.4 Thermo-mechanical analysis

Fleuckiger et al. [101] study the mechanical stresses on the filler material and the tank wall generated by filler relocation due to continuous cycling, so-called thermal ratcheting. They conclude that the internal insulation that is placed between the tank wall and the filler should be increased. They validate their model with experimental measurements in the Solar One storage tank [102].

Gonzáles et al. [103] investigate thermo-mechanical stresses in the tank wall and also validate their model with Solar One data. They obtain high stress levels for thermocline storage tanks with filler material and recommend tank material with high yield stress, caused primarily by hoop stress. With high axial temperature gradients in the tank wall torsion is dominant.

Sassine et al. [104] conduct an analysis of mechanical stresses on the filler and tank. They record highest stresses during loading phases and with increasing bed depth.

Dreißigacker et al. [105] present a model to determine the thermo-mechanical stresses for packed-bed storages with gaseous heat transfer fluids (regenerator storage) and validate it with their own experimental results.

All in all, thermo-mechanical stresses due to thermal cycling lead to mechanical stress in the tank wall and the filler material, so-called thermal ratcheting. For the safe operation of a packed-bed thermocline storage, this has to be considered.

4.2 Experimental studies with conventional fluids

The experimental studies summarized in this section are divided into two parts: thermo-hydraulic measurements and stability tests of filler material.

4.2.1 Thermo-hydraulic measurements

Experimental studies on thermocline storage systems with filler material were initially performed in the first large-scale power tower test plant Solar One in the 1980s with a storage capacity of 170 MWh_{th}. Tests with thermal oil with rocks and sand as filler material (porosity $\varepsilon = 0.22$) successfully proved the potential of dual-media storage systems [38].

Pacheco et al. [39] test a single tank storage (2.3 MWh_{th}) with quartzite rocks and sand as filler material ($\varepsilon = 0.22$) and solar salt as heat transfer fluid. The results are widely used in the literature to validate numerical models.

Mawire and McPherson [106] test a small-scale single tank arrangement, also with a rock/oil mixture ($\varepsilon = 0.42$) with a storage capacity of 2 kWh_{th} and Yang et al. [107] use ceramic spheres and molten salts.

Yin et al. [108] investigate a single tank storage with molten salt only and with two different filler materials: zirconium balls and silicon carbide foam blocks. Highest efficiencies are reached when using no filler material.

Rodat et al. [109] show experimental results from a single-tank storage (2.1 MWh_{th}) applied to a Fresnel plant. They use oil as heat transfer fluid and rocks as filler material. However, the publication gives few thermo-hydraulic results of the storage, as the combination of storage and Fresnel is emphasized.

Bruch et al. [110] present results from pressure and temperature measurements in a lab-scale thermal energy storage unit (0.2 MWh_{th}) with thermal oil as heat transfer fluid and silica rocks and sand as filler material ($\varepsilon = 0.27$). In a subsequent publication, Bruch et al. [111] investigate the influence of repetitive cycling on the performance of a thermocline storage, because, according to the authors, this has not been done sufficiently in the literature yet. In this experimental setup, Esence et al. [112] analyse the start of the charging/discharging from a fully discharged/charged storage.

Hoffmann et al. [113] publish results from a 8.3 kWh_{th} single-tank experiment with rapeseed oil and quartzite rock. They test the influence of two different particle sizes with porosities of 0.41 and 0.39.

In a review by Esence et al. [114], an overview of the main characteristics (tank dimensions, porosity, particle diameter, velocity) of all the mentioned thermo-hydraulic experiments is given. Additionally, experimentally investigated regenerator-type storage systems (gas/solid) are presented.

Odenthal et al. [115, 116] present the new DLR test facility TESIS. A pilot-scale type single tank ($22\,m^3$) is under construction and will be tested with molten salts as heat transfer fluids and low-cost filler material, and temperatures of up to 560 °C.

All in all, experimental data from Pacheco et al. [39] are commonly used for validation purposes. However, these data are scattered and the operational conditions are not clear. Therefore, they are re-testing the single-tank system to gain validation data and operational experience. Only temperatures below 400 °C have been tested to date; the DLR test facility TESIS aims to test molten salts up to 560 °C. No experiments in a single tank with filler material have yet been reported with sodium or any liquid metal as heat transfer fluid.

4.2.2 Stability tests with filler material

Filler material should have low thermal conductivity, but high specific heat capacity and density [117]. Besides the increase of volumetric thermal capacity, the main motivation for using solid filler in a storage system is the material cost reduction and the local abundance. The compatibility of such low-cost filler materials with commonly used heat transfer fluids is tested in the literature.

Brosseau et al. [118] show the stability of quartzite rocks and sand with molten salts in both isothermal and thermal cycling tests of up to 500 °C.

Martin et al. [119] test quartzite and basalt as filler material in contact with solar salt at temperatures up to 560°C.

Grirate et al. [120, 121] investigate six different filler materials abundant in Morocco, in cooperation with CEA. Basalt is the best from a thermal performance point of view; however, it is not suitable for direct contact with thermal oil. Experimental tests show that quartzite and cipolin are best compatible with thermal oil.

Py et al. [122] propose industrial ceramics from asbestos-containing wastes as filler material. In subsequent publications, Calvet et al. [123] and Motte et al. [124] investigate their compatibility with molten salts (up to 500 °C) and air (up to 900 °C). Calvet et al. [125] also propose the use of waste from the steel industry, so-called slag, as low-cost filler material.

John et al. [126] suggest high-temperature concrete as solid material for direct contact with molten salt. They test different concrete mixtures in molten salt baths under isothermal (585 °C) and cyclic conditions. Based on their results, two mixtures are proposed for application in a thermocline storage; however, long term tests are recommended.

All in all, the stability of filler material with thermal oil and molten salt is tested and reported in the literature. Among the tested materials, quartzite proves compatible with the used heat transfer fluids and stable up to 500 °C. Waste products such as ceramics and slag are proposed, as well as locally abundant natural stones.

4.3 Selection of suitable filler material in contact with sodium

No packed-bed thermocline storage system using sodium or any liquid metal has been investigated yet experimentally. Also, no compatibility tests specific to this application have been found in the literature. The literature review on filler material for a packed bed in contact with sodium is split in two parts: Compatible filler material and wettability issues. The costs of the candidate storage filler materials are not included in this section, but will be discussed later in Section 6.1.4.

4.3.1 Compatible filler material

The review is aimed at selecting candidate filler materials for the simulation. Before an experiment is carried out, the compatibility issues should be investigated more deeply. Sodium reacts with water in an exothermic reaction to sodium hydroxide and hydrogen gas, if there is an excess of water, and to sodium monoxide and sodium hydride, if there is an excess of sodium. It also reacts with oxygen to sodium oxide and (under pressure) to sodium superoxide [127]. Besides being compatible with sodium, the solid filler material should have both a high thermal capacity (ρc_p) and relatively low thermal conductivity.

Miller [128] conducted an extensive assessment of the compatibility of numerous materials with liquid metals at three temperature levels: 300 °C, 600 °C and 800 °C. Only the materials with good resistance up to 800 °C are mentioned here. Of the ferrous metals, pure iron, ferritic and austenitic steels show the best compatibility. Among the non-ferrous metals, chromium, nickel and nickel alloys (with Fe,Cr,Mo) resist sodium at high temperatures. As non-metals, only beryllia (BeO) and corundum (Al_2O_3) are suitable. However, beryllia has one of the highest thermal conductivities among the non-metals and it thus not futher considered as a filler material. Fused quartz and silicates show poor corrosion resistance at 800 °C.

A report from EPRI [129] on the compatibility of sodium with insulating materials comes to a similar conclusion. Their tests show that MgO and dense Al_2O_3 are resistant to sodium at 468 °C for 1680 hours.

Stang et al. [130] assess the compatibility of sodium with, among others, steels (austenitic, chromium alloy), nickel- and cobalt-based alloys, graphite and ceramics. Their results basically agree with those of Miller [128]. They show a good resistance of austenitic steels up to 600 °C in contact with sodium. Chromium alloy steels are lower in cost and have a higher conductivity than the austenitic steels. They are as compatible with sodium as austenitic steels, but they are more sensitive to the oxygen content in the sodium. Up to 700 °C, nickel- and cobalt-based alloys are chemically stable with sodium. Ceramics can be compatible with sodium depending on their thermodynamic data, purity and porosity. According to Stang et al. [130], the ceramic carbides (Cr_3C, TiC, ZrC) and Al_2O_3, BeO and Mg_2AlO_4 are chemically stable with sodium at temperatures of up to 800 °C. Graphite, however, is evaluated to have a poor compatibility with sodium. This material combination is assessed in the literature owing to its application in the electrolysis cell in the Hall-Héroult process [131].

Lai [132] reports that carbon steel, Cr-Mo steels and ferritic and austenitic steels show low corrosion rates as temperatures of up to 595 °C. In a publication by Thorley and Bardsley [133] the main reactions between sodium and those steels are described.

Quartzite (SiO_2) is widely investigated in contact with molten salts and thermal oil (as discussed before) due to its physical properties suitable for a packed bed in a thermal storage. According to Götze and Möckel [134], quartzite is a "hard, resistant rock" with a minimum content of 96% quartz (SiO_2). However, the information on compatibility with sodium, especially at high temperatures of up to 700 °C, is scarce. Additionally, there is the effect of quarz inversion at temperatures up to 700 ° undergoing a volume change [135]. Therefore, the silicates need to be heated up carefully. Kikuchi et al. [136] study the kinetics of the reaction of sodium with silica (SiO_2) for application in a fast reactor, as siliceous concrete is widely used in nuclear power plants in Japan. All in all, the information on the compatibility of quartzite with sodium is not sufficient. Nevertheless, it is considered in the simulations in this work to allow for a comparison with the results from molten salts and oil.

Haneefa et al. [127] review the compatibility of concrete with sodium for application in the nuclear industry. They describe the different processes of concrete decomposition in contact with sodium, starting already at 100 °C, mainly with free water and chemically bound water. Casselman [137] also tests the reactivity of concrete with sodium at temperatures of up to 850 °C. The hydrogen release is strongly dependent on the sodium temperature. The main mechanism for decomposition of the concrete is concluded to be by melt soda, which is a product from the water-sodium reaction. Westrich et al. (Sandia National Laboratories) [138] test the compatibility of concrete with sodium at 600 °C and come to the same conclusions. Based on the cited references, concrete is not considered further.

All in all, a good resistance of iron, austenitic steels, chromium alloys, corundum (Al_2O_3) and magnesia spinel (Mg_2AlO_4) in contact with sodium is reported in the literature. The physical properties are listed in the Appendix (Section A.3). The compatibility of quartzite (SiO_2) – which is used in contact with molten salts and thermal oil – with molten sodium is not proven in the literature. Nevertheless, it is used in this work for comparative simulations with molten salt and thermal oil.

4.3.2 Wettability characteristics

According to the Sodium-NaK Engineering Handbook [139], "liquid sodium can be either wetting or nonwetting, depending on temperature or wall material".

There is a critical temperature for each metal, below which wetting with sodium is difficult and above which the wetting by sodium works (see e.g. Figure 4.1). The wettability is dependent on the existence of an oxide film and impurities in the liquid sodium. Experimental works show that the wetting temperature on stainless steel is increased by surface roughness and oxidation and decreased by previous wetting. The wetting characteristics of liquid metals in general is assessed as being complex and the literature shows no consensus, also due to the fact that surface roughness and cleanliness can not be reproduced easily [139].

Bader and Busse [140] experimentally investigate the wetting of sodium on several pure metals, alloys and oxides. Stainless steels (304 L) and corundum (99.7 and 99.5wt% Al_2O_3) are of special interest for this work owing to their good compatibility with sodium at elevated temperatures. The authors investigate the time dependency as well as the temperature dependency. They show that the wetting angle decreases with time, reaching a constant value after a certain interval. The time dependency is due to the reactive nature of sodium. It reacts with the metal oxide film at the wall until equilibrium is reached and the wetting angle stays constant [141]. In the temperature range between 500 and 700 °C no dependency on the temperature was observed [140], however, at lower temperatures there is a high dependency on the temperature with a strongly decreasing wetting angle between 350 and 550 °C (Figure 4.1). A wetting angle of 0–90 degrees means that wetting is possible; a wetting angle of 90–180 degrees means no wetting [142]. Bader and Busse [140] report good wetting angles for all tested materials at temperatures between 520 and 720 °C. For stainless steel, the authors observe wetting angles of 1.7–2.4 degrees (depending on the surface treatment) and 3–7.3 degrees for corundum.

Reed et al. [143] and Viswanathan and Virkar [144] investigate the wetting of sodium on β"-alumina, which is used as the solid electrolyte in Na-S batteries. Viswanathan and Virkar [144] conclude that moisture worsens the wetting and that increasing the temperature (up to 150 °C) leads to improved wetting. Reed et al. [143] report that a tin layer and the removal of surface water increases the wettability at low temperatures (150 °C).

All in all, the cited references make clear that the wettability of materials with sodium is dependent on temperature and surface conditions and that the tests in the literature are not consistent.

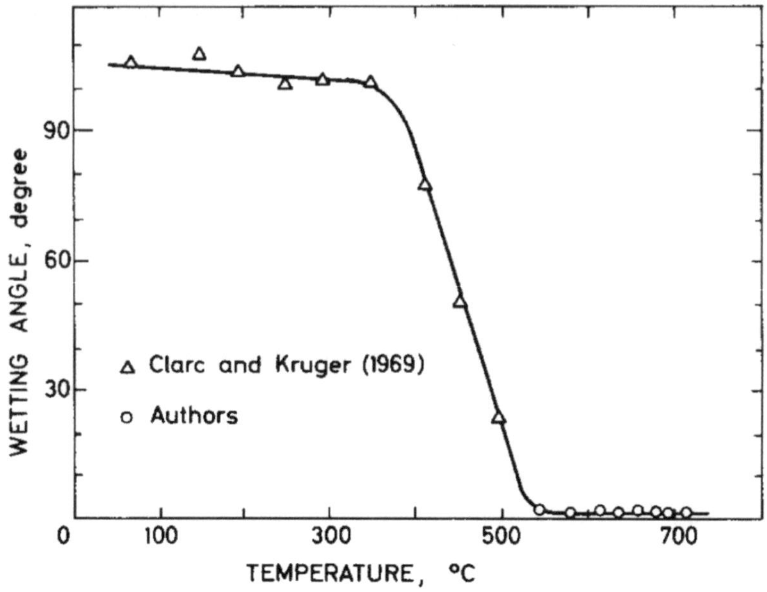

Figure 4.1: Wetting angle of sodium on stainless steel 304 L [140]

The wetting of sodium on two of the selected (due to their compatibility with sodium) materials is experimentally proven to be suitable above temperatures of $\approx 500\,°C$: stainless steel and corundum.

4.4 Conclusion

This section presents the current state of research concerning thermocline packed-bed storage systems. Previous theoretical studies show that the parameters tank dimensions, porosity and filler dimensions have the greatest influence on packed-bed thermocline storage systems with conventional heat transfer fluids. Therefore, those will also be investigated for the sodium storage system. Furthermore, different fluids (molten salts, oils and air) are compared in the literature; however, there is no consensus about the definition of the efficiency of such a thermal energy storage system and pumping power, and storage costs are typically not considered.

Experimental results are scarce. Usually, the experimental data of a pilot-scale experiment in the Sandia National Laboratories [39] with molten salt and rocks/sand are used for validation purposes. Data from a large-scale experiment as part of the test plant Solar One are also available (oil with rocks/sand). A few further lab-scale experiments have been conducted with molten salts and thermal oil as heat transfer fluids.

For sodium, neither an experimental nor a systematic theroretical study of a packed-bed thermocline storage has yet been reported. Thus, this work aims to theoretically ascertain if this configuration is feasible with molten sodium and to determine the optimal storage parameters. It will be a basis for the design of an experimental setup. For the simulation of the storage system with molten sodium, filler materials that are compatible with sodium are selected: pure iron, austenitic steel, Al_2O_3, Mg_2AlO_4 and quartzite (for comparative reasons). The chemical compatibility of the filler materials with sodium, especially at temperatures beyond 600 °C, the cycling and wetting behaviour need to be assessed before an experiment is conducted.

5 Modelling of a single-tank packed-bed storage with sodium

In this section, the numerical models used to calculate the temperature distributions in a packed-bed storage with sodium as heat transfer fluid is presented. As two different models are used for cycling (charging/discharging) and standby, this section is divided based on these two operational modes.

5.1 Cycling

A cycle consists of charging and discharging (Figure 5.1).

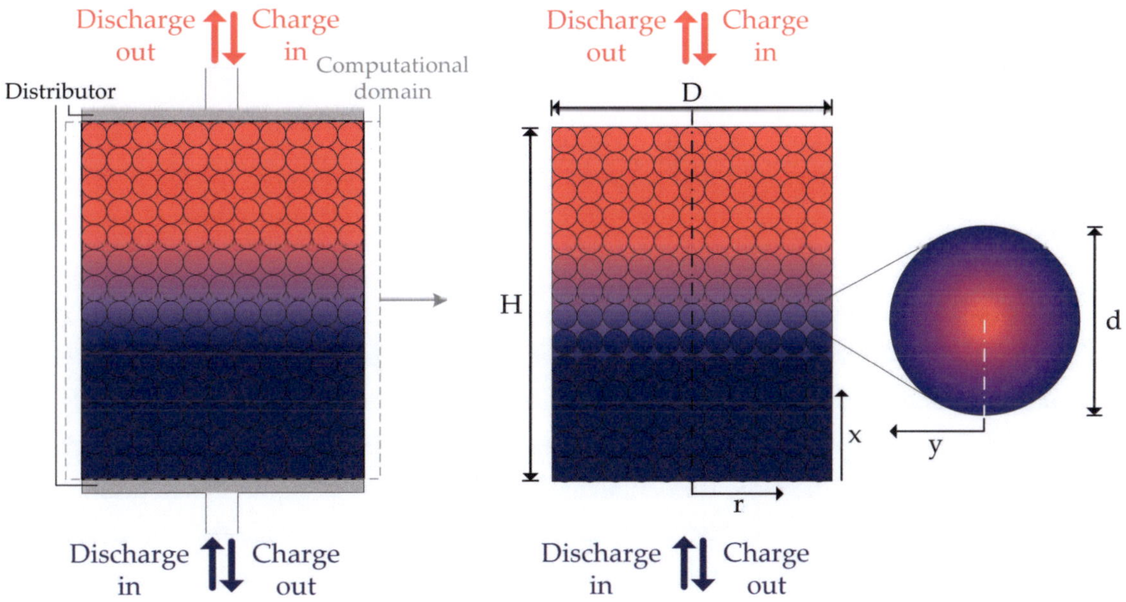

Figure 5.1: Principle of discharging with cold fluid (from the bottom) and charging with hot fluid (from the top); left: including distributors; right: computational domain

During charging hot fluid enters the tank at the top and thermal energy is transferred to the cold filler material. During discharging cold fluid enters from the bottom of the tank and thermal energy is transferred from the hot filler material to the fluid.

In the following, the governing energy equations for cycling are presented (Section 5.1.1), and a two-dimensional and one-dimensional model are compared (Section 5.1.2). Further, the equation parameters are discussed and adapted to liquid metal (Section 5.1.3). Finally, the numerical solution procedure and implementation in MATLAB (Section 5.1.4) and the validation of the code (Section 5.1.5) are shown.

5.1.1 Governing equations

To calculate the temperature distribution in the packed-bed storage, different models are proposed in the literature [114, 145]. They are listed and discussed in the Appendix (Section A.5.1). A two-phase two-dimensional (2P-2D) model including intra-particle diffusion is determined as suitable for this study in order to calculate the temperature distribution in the packed-bed thermocline storage system with sodium as heat transfer fluid. For salt, a simplified 2P-2D model assuming lumped capacitance of the solid phase can be used.

Governing equations of the 2P-2D model. The temperature distribution in the fluid phase and one representative particle are calculated by solving Equations 5.1 and 5.2.

$$\varepsilon \rho_{\mathrm{f}} c_{p\mathrm{f}} \left(\frac{\partial T}{\partial t} + u \frac{\partial T}{\partial x} \right) = k_{\mathrm{fx}} \frac{\partial^2 T}{\partial x^2} + k_{\mathrm{fr}} \left(\frac{\partial^2 T}{\partial r^2} + \frac{1}{r} \frac{\partial T}{\partial r} \right) - h_{\mathrm{v}} (T - T_{\mathrm{p}}|_{y=\frac{d}{2}}) \tag{5.1}$$

$$\rho_{\mathrm{s}} c_{p\mathrm{s}} \frac{\partial T_{\mathrm{p}}}{\partial t} = \lambda_{\mathrm{s}} \left(\frac{\partial^2 T_{\mathrm{p}}}{\partial y^2} + \frac{2}{y} \frac{\partial T_{\mathrm{p}}}{\partial y} \right) \tag{5.2}$$

On the left-hand side of the fluid equation are the enthalpy and accumulation terms, on the right side the thermal conduction in axial and radial direction and the heat transfer to the solid phase. Heat losses are considered in the boundary condition at the tank wall (Section A.5.4). The energy equations of the fluid and particle are coupled by the heat transfer term of the fluid phase on the one hand and the boundary condition at the outer surface of the filler particles (Section A.5.4) on the other hand. Owing to the large Biot numbers (Bi $\gg$ 1) in case of liquid metal, meaning that the heat transfer resistance in the filler material is significantly higher than the one at the outer surface, the radial temperature distribution in the particles needs to be considered. The particles are assumed as ideal spheres with a concentric thermal distribution in the particles and a uniform fluid temperature around each particle (Figure 5.2a).

The solid spheres are expected to have only point contact and a significantly lower thermal conductivity than the fluid in case of liquid metal as heat transfer fluid. Therefore, axial and radial thermal diffusion in the solid phase is neglected in this work. Wakao and Kaguei [146] present an approach to include the solid conduction in the effective thermal conductivity of the fluid (Figure 5.2b), which is, however, determined from an experiment with air as heat transfer fluid, where

Figure 5.2: Schematic heat transfer models, a) fluid: dispersed plug flow, solid: concentric isotherms, b) modified effective thermal conductivity of fluid including solid phase (adapted from Ref. [146])

increased axial effective thermal conductivites were observed, as compared with air alone. Radiation effects are neglected in this work, as this can be assumed for liquids in porous systems according to Kunii and Smith [147]. Natural convection effects are also neglected, as, according to Yang et al. [107], the filler material limits this phenomenon in the fluid. In the Appendix (Section A.5.2) the natural convection is estimated.

For low Biot numbers (Bi < 1), e.g. molten salts or thermal oil, which will be used in comparative studies, the energy conservation equations for the fluid and homogeneous solid phase are shown in Equation 5.3 and Equation 5.4, respectively.

$$\varepsilon \rho_{\mathrm{f}} c_{p\mathrm{f}} \left(\frac{\partial T}{\partial t} + u \frac{\partial T}{\partial x} \right) = k_{\mathrm{fx}} \frac{\partial^2 T}{\partial x^2} + k_{\mathrm{fr}} \left(\frac{\partial^2 T}{\partial r^2} + \frac{1}{r} \frac{\partial T}{\partial r} \right) - h_{\mathrm{v}} (T - T_{\mathrm{s}}) \tag{5.3}$$

$$(1-\varepsilon) \rho_{\mathrm{s}} c_{p\mathrm{s}} \frac{\partial T_{\mathrm{s}}}{\partial t} = k_{\mathrm{sx}} \frac{\partial^2 T_{\mathrm{s}}}{\partial x^2} + k_{\mathrm{sr}} \left(\frac{\partial^2 T_{\mathrm{s}}}{\partial r^2} + \frac{1}{r} \frac{\partial T_{\mathrm{s}}}{\partial r} \right) + h_{\mathrm{v}} (T - T_{\mathrm{s}}) \tag{5.4}$$

The energy equation for the fluid consists of the same terms as described above; however, now the equations are directly coupled by the heat transfer term. The energy equation of the solid phase is composed of the same terms as of the fluid phase except for the enthalpy term, as the solid does not move (Figure 5.3).

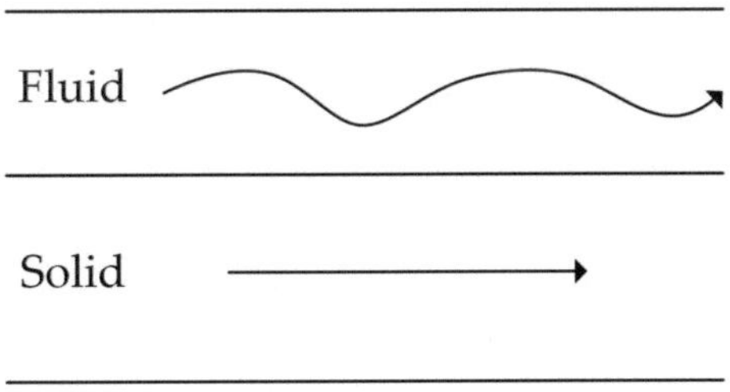

Figure 5.3: Schematic heat transfer model for Bi < 1, fluid: dispersed plug flow, solid: axial heat conduction; adapted from Ref. [146]

In the literature, the following simplifications are commonly applied to the solid phase. They can also be employed in this work, as those are not specific for the heat transfer fluid used.

- Heat losses occur only from fluid to ambient due to point contact of the filler particles at the tank wall.
- Thermal conduction (radial and axial) in the solid phase can be neglected as compared with the fluid phase, owing to point contact of the filler particles.

This simplifies the energy equation of the solid phase (for the lumped capacitance model) to Equation 5.5. In this form, it can be solved analytically, which simplifies the solution procedure considerably. A comparison of the results for including and neglecting the axial heat conduction in the solid phase is shown in the Appendix (Section A.5.3).

$$(1-\varepsilon)\rho_{\mathrm{s}}c_{p\mathrm{s}}\frac{\partial T_{\mathrm{s}}}{\partial t} = h_{\mathrm{v}}(T-T_{\mathrm{s}}) \tag{5.5}$$

The initial and boundary conditions are defined in the Appendix (Section A.5.4).

Governing equations of the simplified 2P-1D model. The energy equation of the fluid phase can be further simplified by neglecting temperature gradients in the radial tank axis. Now, eventual heat losses are applied to the overall volume and not in the boundary condition. The energy equations of the particles and solid (lumped capacitance) remain unchanged.

For all Biot numbers:

$$\varepsilon\rho_{\mathrm{f}}c_{p\mathrm{f}}\left(\frac{\partial T}{\partial t}+u\frac{\partial T}{\partial x}\right) = k_{\mathrm{fx}}\frac{\partial^2 T}{\partial x^2} - h_{\mathrm{v}}(T-T_{\mathrm{p}}\big|_{y=\frac{d}{2}}) - h_{\mathrm{w,v}}(T-T_0) \tag{5.6}$$

For Bi < 1:

$$\varepsilon\rho_{\mathrm{f}}c_{p\mathrm{f}}\left(\frac{\partial T}{\partial t}+u\frac{\partial T}{\partial x}\right) = k_{\mathrm{fx}}\frac{\partial^2 T}{\partial x^2} - h_{\mathrm{v}}(T-T_{\mathrm{s}}) - h_{\mathrm{w,v}}(T-T_0) \tag{5.7}$$

For the parametric studies and comparative studies with conventional heat transfer fluids an adiabatic storage tank is assumed and the heat transfer coefficient to the ambient is set to zero: $h_{\mathrm{w,v}} = 0$.

5.1.2 Two-dimensional vs. one-dimensional model

The results from the two-phase two-dimensional (2D) and one-dimensional (1D) model are compared by applying different heat loss cases: adiabatic, insulated and non-insulated. The heat transfer coefficients at the tank wall for each case are given in Section 5.1.3. The storage tank of Sandia National Laboratories [39] is taken as reference, with the heat transfer fluid being replaced by sodium. The temperature distributions after 1 h of discharging are compared as an example. Figure 5.4 shows that for adiabatic and insulated tanks it is sufficient to solve the simplified 1D model without considering radial effects.

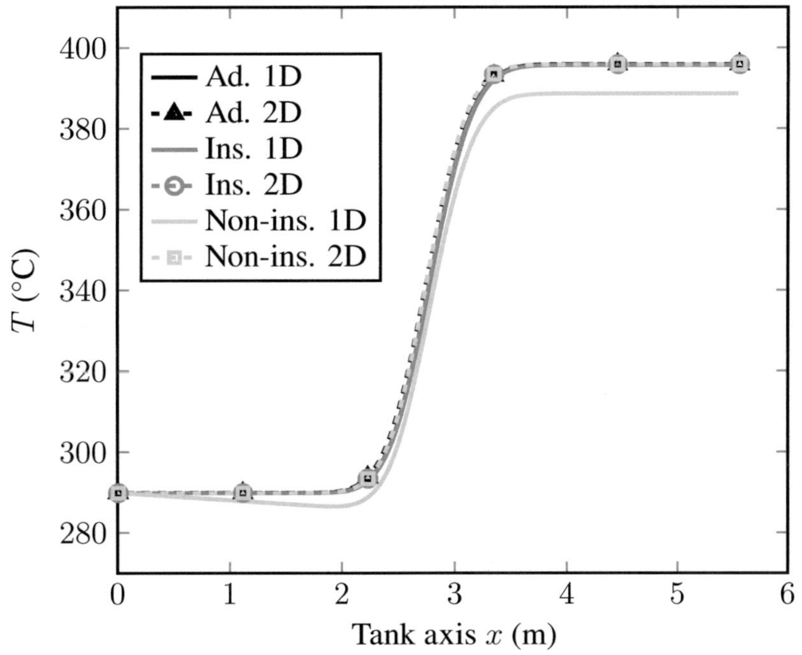

Figure 5.4: Comparison of temperature distribution during discharge (after 1 h) resulting from different assumed heat losses (ad. = adiabatic, ins. = insulated), 1D and 2D (at $r = 0$) two-phase models

For non-insulated tanks (h_w = 10 W/(m^{2}K)) there are differences in the results of both models. With the 1D model, the heat losses to the ambient influence the temperature distribution in the tank, especially in the hotter upper part of the storage. The least influence is obtained at the entry of the tank due to the constant temperature boundary condition where the cold fluid enters. With the 2D model, by contrast, no influence on the temperature distribution in the centre of the tank ($r = 0$) is visible at the given fluid velocity. The heat transfer in the tank (from solid to fluid) and the thermal conductivity in axial direction dominate the discharging process. The heat losses play an insignificant role, as they are considered only at the outer wall and not being subtracted from the overall volume energy as it is the case in the 1D model. Only with significantly lower velocities (e.g. a overall discharging time of 10 h instead of 2 h) is an influence of the heat losses on the temperature distribution in the centre of the tank observable.

The comparison of the results from the 1D and the 2D model shows that for the cycling process with an adiabatic or well-insulated tank, a 1D model is sufficient and leads to results similar to those of the 2D model.

5.1.3 Parameters ε, u, k_f, h_v and h_w

In this section, the equation parameters ε, u, k_f, h_v and h_w are given and discussed[2]. In particular, adaptations and assumptions for liquid metal are presented.

Bed porosity ε. The porosity ε is a volume porosity and it is defined by the empty volume, which is filled by the fluid, divided by the total volume of the tank [148].

$$\varepsilon = \frac{V_\mathrm{f}}{V_\mathrm{tot}} = \frac{V_\mathrm{tot} - V_\mathrm{s}}{V_\mathrm{tot}} \tag{5.8}$$

The minimum porosity of a packed bed of monodisperse spheres is 0.26 in the face-centered cubic system. The usual porosity of a packed bed with monodisperse spherical particles is 0.38–0.41 [148]. Lower porosities can be achieved in bidisperse packed beds, as discussed in Section 6.3.1. The porosity depends on the tank radius with a maxium of unity at the tank wall if point contact is assumed. After a few particle diameters, the porosity converges to the bed porosity. For large tank to particle diameter (D/d) ratios, which is the case in this work, the influence of the wall effect on the flow pattern and heat transfer is negligible [148]. In Figure 5.5 a schematic porosity distribution in a packed bed of sperical particles along the wall distance as a multiple z of the particle diameter d is shown. In this case, the bed porosity does not change significantly after a distance of five filler diameters.

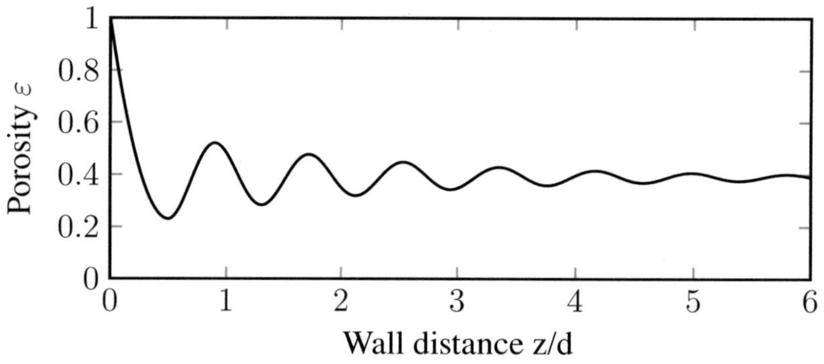

Figure 5.5: Local porosity in packed beds of spherical particles using the equations and coefficients ($\varepsilon_\mathrm{min} = 0.23$ (at $z = d/2$), $\varepsilon_\mathrm{bed} = 0.39$ and the period of the oscillation $T = 0.816$) as given in Ref. [148]

Fluid velocity u. The fluid velocity u is assumed to be uniform across the cross section of the tank. Indeed, such tanks will be equipped with well-designed distributors, so that this assumption

[2] Preliminary work regarding the definition of k_f, h_v and h_w is performed within the framework in the Bachelor thesis of J. Kleinheins (2017).

is acceptable for the simulation. The fluid velocity u is defined in the following with the empty part of the cross section ($\pi \frac{D^2}{4} \cdot \varepsilon$) [148].

$$u = \frac{u_0}{\varepsilon} = \frac{4\dot{m}}{\rho_f \pi D^2 \varepsilon} \tag{5.9}$$

According to Schlünder and Tsotsas [148], the velocity u_0 increases with increasing radial coordinate, has a maximum and slows down at the tank wall. This radial dependence of the velocity is negligible for high values of the D/d ratio, which are used in this work. In Ref. [148] an empirical equation is given to calculate the increase of the ratio of the velocity u_0 normalized with the mean value $\bar{u}_0$ (Equation 5.10), the coefficients K and P are functions of D/d. However, the decrease close to the tank wall is neglected in this equation.

$$\frac{u_0}{\bar{u}_0} = \frac{K + \left(\frac{r}{R}\right)^P \frac{P+2}{2}}{K+1} \tag{5.10}$$

Figure 5.6 shows a schematic velocity profile along the radial coordinate r/R with a ratio of $D/d = 5.8/0.015 = 387$, which is representative of the majority of simulations done in this work. The minimum D/d figure used in this work is $D/d = 5.8/0.05 = 156$. Even in this case, the deviation from the mean velocity is less than 0.15%.

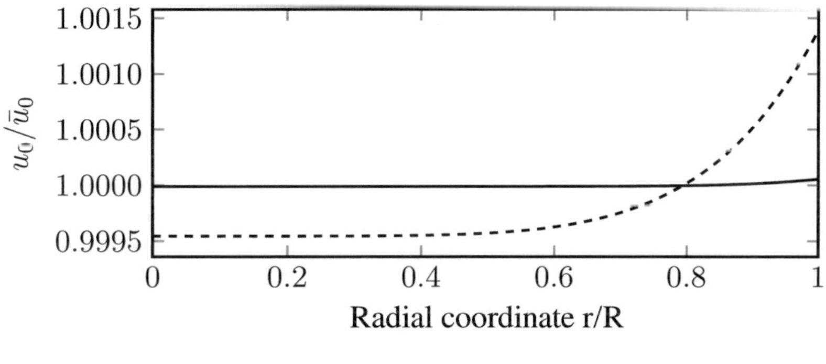

Figure 5.6: Local velocity in packed beds of spherical particles using the equations and coefficients as given in Ref. [148]

Effective thermal conductivity k_f. The effective thermal conductivity in axial direction k_{fx} and in radial direction k_{fr} are defined in the following. First of all, an isotropic porous structure is assumed in axial and radial direction:

$$k_{fx} = k_{fr} = k_f \tag{5.11}$$

In the literature, a variety of definitions of the effective heat conductivity of the fluid k_f can be found, as exemplified in Table 5.1. The discrepancy is often based on the idea that this parameter not only represents the thermal conductivity of the fluid, but should also include a contribution of the thermal conductivity in the solid phase and additional heat transfer mechanisms such as radiation and natural convection. Especially for one-phase (homogeneous) models this is usually done, as the heat transfer mechanisms cannot be considered in the respective phases [145].

Table 5.1: Definitions of effective conductivity k_f used in the literature in two-phase models

$k_f =$		Ref.
$k_{f1} = \varepsilon \lambda_f$	(5.12)	[91, 117, 149, 150]
$k_{f2} = \begin{cases} 0.7\varepsilon\lambda_f(\mathrm{Re_p} \leq 0.8) \\ 0.5\mathrm{PrRe_p}\lambda_f(\mathrm{Re_p} > 0.8) \end{cases}$	(5.13)	[90, 145, 151]
$k_{f3} = f(\lambda_f, \beta, \phi)$ with $\phi = 1-\varepsilon$; $\beta = \frac{\lambda_s-\lambda_f}{\lambda_s+2\lambda_f}$	(5.14)	[85]
$k_{f4} = k_{bed} + 0.5\mathrm{PrRe_p}\lambda_f$ with $k_{bed} = k_{ZBS}$ (Section 5.2.1)	(5.15)	[152]

Xu et al. [153] compare effective conductivities for a molten salt packed bed using, among others, the definitions k_{f1} to k_{f3} given in Table 5.1. They state that the calculated fluid temperatures (for their system) do not differ significantly from each other; only with k_{f2} is a slightly thicker thermocline region reached. However, it is not clear in which range of Prandtl numbers this correlation is valid. The third definition (k_{f3}) is valid for porosities of $0.15-0.85$ and β from -0.499 to 1 [154] and is based on the analytical solution of Maxwell considering the overall effective conductivity in a packed bed. It is again suitable for a homogeneous (one-phase) model.

Tsotsas [152] defines the axial dispersion coefficient k_{f4} in an axial flow packed bed as a sum of the effective thermal diffusivity in the packed bed without a flow k_{bed} and the molecular Peclet number ($\mathrm{Pe_0} = \mathrm{PrRe_p}$). The effective thermal diffusivity in the packed bed without a flow k_{bed} is calculated according to Zehner, Bauer and Schlünder, as explained in Section 5.2.1.

In Figure 5.7 the temperature distributions simulated with the four definitions of k_f are compared. The storage tank of Sandia National Laboratories [39] is taken as a reference with the heat transfer fluid being replaced by sodium.

The temperature distributions do not show large differences. In general, the higher the effective thermal conductivity of the fluid phase, the flatter the slope of the fluid temperature along the tank axis. In this work, the widely applied k_{f1} is used for the effective conductivity in the fluid phase. However, it is noted that, as there is an inconsistency in the literature regarding the axial effective thermal conductivity in packed beds, this should be investigated deeper with suitable experiments.

Volumetric heat transfer coefficient between fluid and solid phase h_v. The volumetric heat transfer coefficient h_v is the parameter that couples the fluid and solid/particle equations. It is calculated from the specific surface s_v and the (surface specific) heat transfer coefficient α.

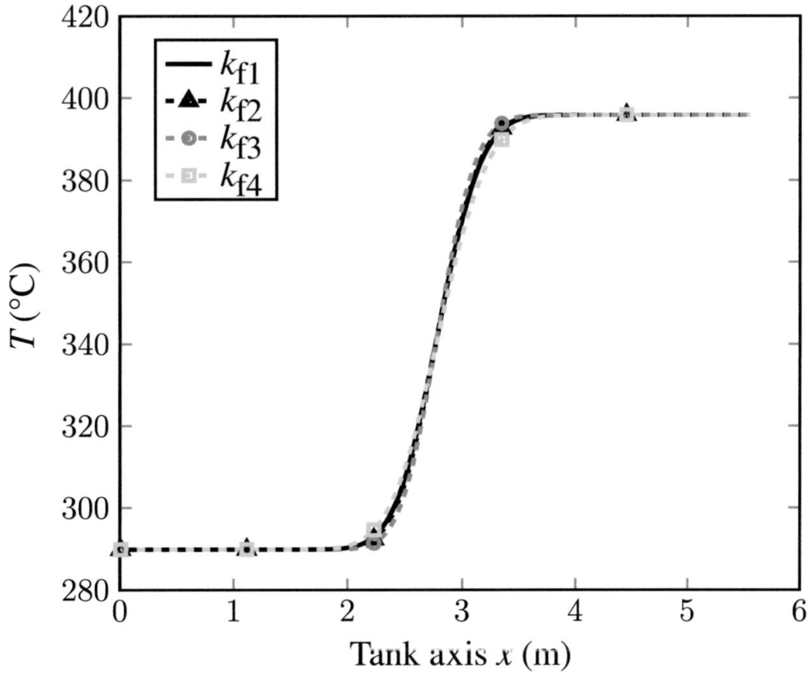

Figure 5.7: Fluid temperature distribution in sodium/quartzite packed bed along tank axis after 1 h of discharge resulting from different k_f (at T_{max}); k_{f1} = 15.9 W/(mK), k_{f2} = 13.9 W/(mK), k_{f3} = 8.5 W/(mK), k_{f4} = 25.0 W/(mK)

$$h_v = s_v \alpha \tag{5.16}$$

The specific surface is calculated from the surface of all spherical particles $A_{p,tot}$ in a monodisperse packed bed with uniform porosity relative to the overall volume of the storage tank V_{tot} [155].

$$s_v = \frac{A_{p,tot}}{V_{tot}} = \frac{A_{p,tot}}{V_{p,tot}} \frac{V_{p,tot}}{V_{tot}} = \frac{\pi d^2}{\frac{1}{6}\pi d^3}(1-\varepsilon) = \frac{6(1-\varepsilon)}{d} \tag{5.17}$$

The heat transfer coefficient α is determined by the Nusselt number for the packed bed $\mathrm{Nu_{bed}}$, the thermal coductivity of the fluid λ_f and the characteristic length of a sphere being its diameter d [156].

$$\alpha = \frac{\mathrm{Nu_{bed}} \cdot \lambda_f}{d} \tag{5.18}$$

For conventional fluids (molten salts and thermal oil), the Nusselt correlation by Wakao and Kaguei [146] is used in the majority of the publications. Strictly speaking, it is valid only for gases (Pr $\approx$ 0.7). This correlation is used in this work for validation and comparison with molten salt and thermal oil:

$$\mathrm{Nu_{bed,[146]}} = 2 + 1.1\mathrm{Re_p^{0.6}Pr^{1/3}} \qquad 15 \leq \mathrm{Re_p} \leq 8500 \tag{5.19}$$

Xu et al. [153] compare the above and four other Nusselt correlations for molten salt, with the result that no significant differences in the resulting temperatures are obtainable.

For the low Prandtl number fluid liquid metal, however, no Nusselt correlation for a packed bed is available in the literature. Melissari and Argyropoulos [157] propose an equation for a wide range of Prandtl numbers (from sodium to water: $0.003 < \mathrm{Pr} < 10$), but only for the heat transfer to a single sphere. The correlation is obtained from a model of a melting liquid metal sphere and validated with two experiments of melting a magnesium-alloy and an aluminium sphere.

$$\mathrm{Nu}_{[157]} = 2 + 0.47\mathrm{Re}_\mathrm{p}^{0.5}\mathrm{Pr}^{0.36} \qquad 10^2 \leq \mathrm{Re}_\mathrm{p} \leq 5 \cdot 10^4 \tag{5.20}$$

Gnielinski [158] presents a relation between the Nusselt correlation of a single sphere and a packed bed in terms of a shape factor f_a, which is given in Equation 5.22 for a monodisperse packed bed of spheres.

$$\mathrm{Nu}_{\mathrm{bed},[158]} = \mathrm{Nu} \cdot f_\mathrm{a} \tag{5.21}$$

$$f_\mathrm{a} = 1 + 1.5(1 - \varepsilon) \tag{5.22}$$

However, this correlation is valid only for porosities $\varepsilon > 0.26$ and Prandtl numbers > 0.7. Therefore, the combination of the Nusselt correlation (Equation 5.20) and the shape factor is not chosen in this work, as Equation 5.21 is not valid for the Prandtl number range of sodium, and in some operational cases also the Reynolds number range is below the valid range of Equation 5.20. In the absence of a valid Nusselt correlation for sodium, only conductive heat transfer is considered, leading to the minimum Nusselt number for a sphere: Nu = 2.

$$\mathrm{Nu} = \frac{\alpha d}{\lambda_\mathrm{f}} = \frac{\frac{\lambda_\mathrm{f}}{d/2} d}{\lambda_\mathrm{f}} = 2 \tag{5.23}$$

In most cases, the Nusselt number would be close to the minimum number in a packed-bed storage, as the Reynolds number is very low due to very slow fluid flow ($\approx$ 1 mm/s) and small filler particles ($\approx$ 10 mm) and the low Prandtl number of liquid sodium ($\approx$ 0.005).

In Figure 5.8 the influence of an increased Nusselt number on the fluid temperature distributions along the tank axis are shown. Again, the packed-bed storage system of Sandia National Laboratories [39] is taken as reference with the fluid being replaced by sodium. There is no large influence on the fluid temperature distribution when increasing the Nusselt number from 2 to 10. In general, the larger the the Nusselt number the better the heat transfer to the filler particles and thus, the sharper the slope of the curve and the smaller the thermocline region. Therefore, taking Nu = 2 in the absence of a suitable correlation is acceptable.

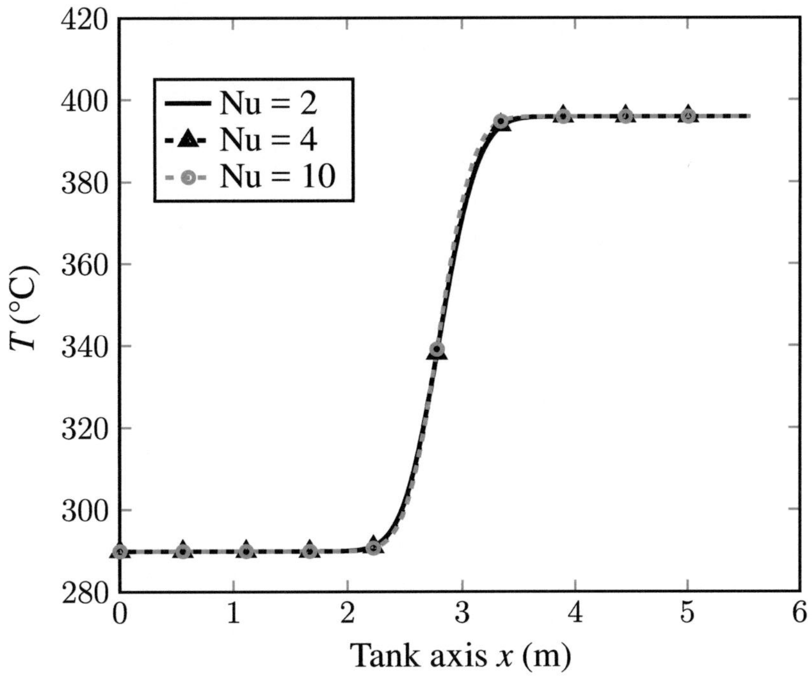

Figure 5.8: Fluid temperature distribution in sodium/quartzite packed bed along tank axis after 1 h of discharge resulting from different Nusselt numbers

By combining Equations 5.16, 5.17, 5.18 and 5.23, the volumetric heat transfer coefficient h_v used in this work for liquid metal is defined in the following.

$$h_\mathrm{v} = \frac{6(1-\varepsilon)\cdot 2\lambda_\mathrm{f}}{d^2} \tag{5.24}$$

Heat transfer coefficient through tank wall h_w. The heat transfer coefficient at the tank wall is used to apply realistic heat losses and assess the effects in the radial temperature distribution. Heat losses are applied only to the tank wall; the bottom and roof of the tank are assumed to be ideally insulated and not affected by heat losses.

The heat transfer occurs through the tank wall and through thermal insulation layers. As tank wall material stainless steel is used, as suggested in Refs. [99, 101, 153]. Flueckiger et al. [101] propose a further inner insulation consisting of firebrick, which is protected by a thin liner. The latter can be neglected in the thermal assessment. As an outer layer ceramic is proposed for insulation and corrosion protection.

In this work, the material combination of Flueckiger et al. [101] is assumed and schematically shown in Figure 5.9.

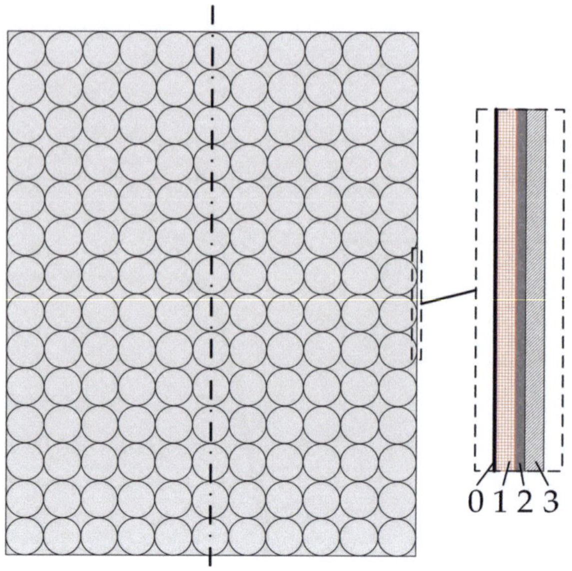

Figure 5.9: Scheme of the tank wall and insulation, consisting of thin liner *(0)*, firebrick insulation *(1)*, steel wall *(2)* and ceramic insulation *(3)*; adapted from Ref. [101]

The properties and thicknesses of the different layers are taken from Ref. [153] and listed in Table 5.2. The thermal energy storage system investigated by Xu and al. [153] is of a capacity similar to the one selected in this work (Section 6.2) and at lower temperatures ($\approx 350\,°\mathrm{C}$); thus, these layers are also used in the present work.

Table 5.2: Tank wall and insulation thicknesses and thermal conductivities taken from Ref. [153]

Index i	Material	Thickness s_i (m)	λ_i (W/(mK))
1	Firebrick	0.2	0.1
2	Stainless steel	0.04	35
3	Ceramic	0.2	0.1

The parameter h_w is an overall heat transfer coefficient and is calculated according to Ref. [159]. As the tank diameter is significantly larger than the tank wall thickness, the heat transfer coefficient h^w can be calculated as for a plane wall:

$$\frac{1}{h_\mathrm{w}} = \frac{1}{\alpha_\mathrm{inner}} + \sum_{n=1}^{3} \frac{s_i}{\lambda_i} + \frac{1}{\alpha_\mathrm{outer}} \tag{5.25}$$

Due to the large thermal conductivity of liquid metals, it can be assumed that the wall temperature is similar to the fluid temperature and thus, the inner resistance can be neglected ($\frac{1}{\alpha_\mathrm{inner}} \approx 0$). The outer heat transfer coefficient is calculated from forced convection, which dominates the heat transfer when an air velocity of 5 m/s is assumed, across a cylinder with the flow length $l = \pi \cdot D/2$ according to Ref. [160].

$$\alpha_{\text{outer}} = \frac{\text{Nu}_{\text{l}} \lambda_{\text{air}}}{l} \tag{5.26}$$

$$\text{Nu}_{\text{l}} = 0.3 + \sqrt{\text{Nu}_{\text{l,lam}}^2 + \text{Nu}_{\text{l,turb}}^2} \tag{5.27}$$

$$\text{Nu}_{\text{l,lam}} = 0.664 \sqrt{\text{Re}_{\text{l}}} \sqrt[3]{\text{Pr}} \tag{5.28}$$

$$\text{Nu}_{\text{l,turb}} = \frac{0.037\,\text{Re}_{\text{l}}^{0.8} \text{Pr}}{1 + 2.443\,\text{Re}_{\text{l}}^{-0.1} (\text{Pr}^{2/3} - 1)} \tag{5.29}$$

The Reynolds number is defined with the flow length l across the cylinder. For this estimation, the diameter of the cylinder (storage tank) D is assumed to be 10 m. The properties of air are taken from Ref. [161]. This leads to $\text{Nu}_{\text{forced}} \approx 6975$ and α_{outer} = 11.5 W/(m^{2}K). In the literature, a constant value of 5–10 W/(m^{2}K) is usually assumed [101]. In this work, an outer heat transfer coefficient of α_{outer} = 10 W/(m^{2}K) is used.

When the figures from Table 5.2 and α_{outer} are inserted in Equation 5.25, the overall heat transfer coefficient results in $h_{\text{w,ins.}}$ = 0.24 W/(m^{2}K). For the simulation of non-insulated tanks, only the steel tank wall without insulation is considered, resulting in $h_{\text{w,non-ins.}} = \left(\frac{s_2}{\lambda_2} + \frac{1}{\alpha_{\text{outer}}}\right)^{-1} =$ 9.89 W/(m^{2}K).

This parameter is used in the boundary condition at the tank wall when a two-dimensional model is used. When the simplified one-dimensional model is used, the term is included in the fluid energy equation as a volumetric heat loss term. Therefore, h_{w} is multiplied with the specific surface of the tank wall.

$$h_{\text{w,v}} = s_{\text{v,w}} h_{\text{w}} \tag{5.30}$$

$$s_{\text{v,w}} = \frac{A_{\text{w}}}{V_{\text{tot}}} = \frac{\pi D L}{\pi (D^2/4) L} = \frac{4}{D} \tag{5.31}$$

Table 5.3 summarizes the definitions of the parameters of the energy equations that are solved in the present work and the involved assumptions discussed in this section.

For liquid metal and conventional fluids (molten salt, thermal oil) the same definitions are used, with the exception of the Nusselt correlation for the heat transfer between fluid and filler. For molten salt, a correlation of Wakao and Kaguei [146] is used, whereas for liquid metals a minimum Nusselt number is used due to the lack of a valid Nusselt correlation for low Prandtl number fluids.

Table 5.3: Overview of parameters of the energy equations used in the present work for liquid metals

Parameter	ε	u	k_f	h_v	h_w
Definition	$\frac{V_f}{V_{tot}}$	$\frac{u_0}{\varepsilon}$	$\varepsilon\lambda_f$	$\frac{6(1-\varepsilon)\cdot 2\lambda_f}{d^2}$	$\left(\sum_{n=1}^{3}\frac{s_i}{\lambda_i}+\frac{1}{\alpha_{outer}}\right)^{-1}$
Assumptions	constant along tank radius due to negligible wall effects for large D/d	constant along tank radius due to negligible radial effects for large D/d	no inclusion of thermal conductivity of solid phase and other heat transfer mechanisms	only conductive heat transfer in absence of Nusselt correlation	$\alpha_{inner} = 0$, $\alpha_{outer} = 10\,W/(m^2K)$, insulation layers according to Ref. [101]

5.1.4 Solution procedure and implementation in MATLAB

The solution of the differential equations is performed with the finite volume method and coded in MATLAB[3].

Finite volume method. The finite volume (FV) method uses the integral form of the conservation equation. The domain is subdivided into a finite number of control volumes (CVs). A suitable number of CVs is determined and the calculation is then performed in the centre of each control volume. This method is described in detail in Refs. [162, 163]. For solving the equation in each node, the volume and surface integrals are numerically approximated by using differencing schemes. The notation used typically is shown in Figure 5.10. The characters "e", "w", "n", "s" stand for their orientation, which is east, west, north or south of the centre P. Small letters refer to positions at the wall, capital letters to the centre of a node.

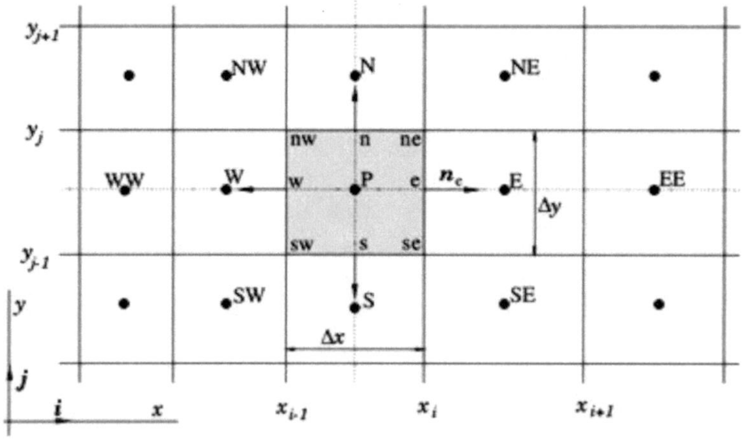

Figure 5.10: Location and notation of central node and surrounding nodes [162]

The energy equation terms are integrated for the internal nodes, therefore the heat-loss term is not considered here, but in the boundary nodes. The tank is considered to be in a vertical position

[3] The numerical implementation of the differential equation system in MATLAB has been done by L. Marocco (Politecnico di Milano).

with the fluid always entering south and exiting north. The fluid equation is integrated for each node over the control volume and time. The integration of the energy equations for the fluid and solid phase is shown in the Appendix (Section A.5.5) in detail, as well as the analytical solution for the energy equation of the solid phase (Equation 5.5) if a lumped capacitance is assumed. In this work, a first-type upwind differencing scheme (UDS) is used for the spatial discretization of the advective terms and a central differencing scheme (CDS) for the diffusive terms. A Crank-Nicolson scheme is used for time discretization.

Implementation in MATLAB. A main function (*main.m*) calls all sub-functions in which the equations are solved. The schematic structure is presented in Figure 5.11.

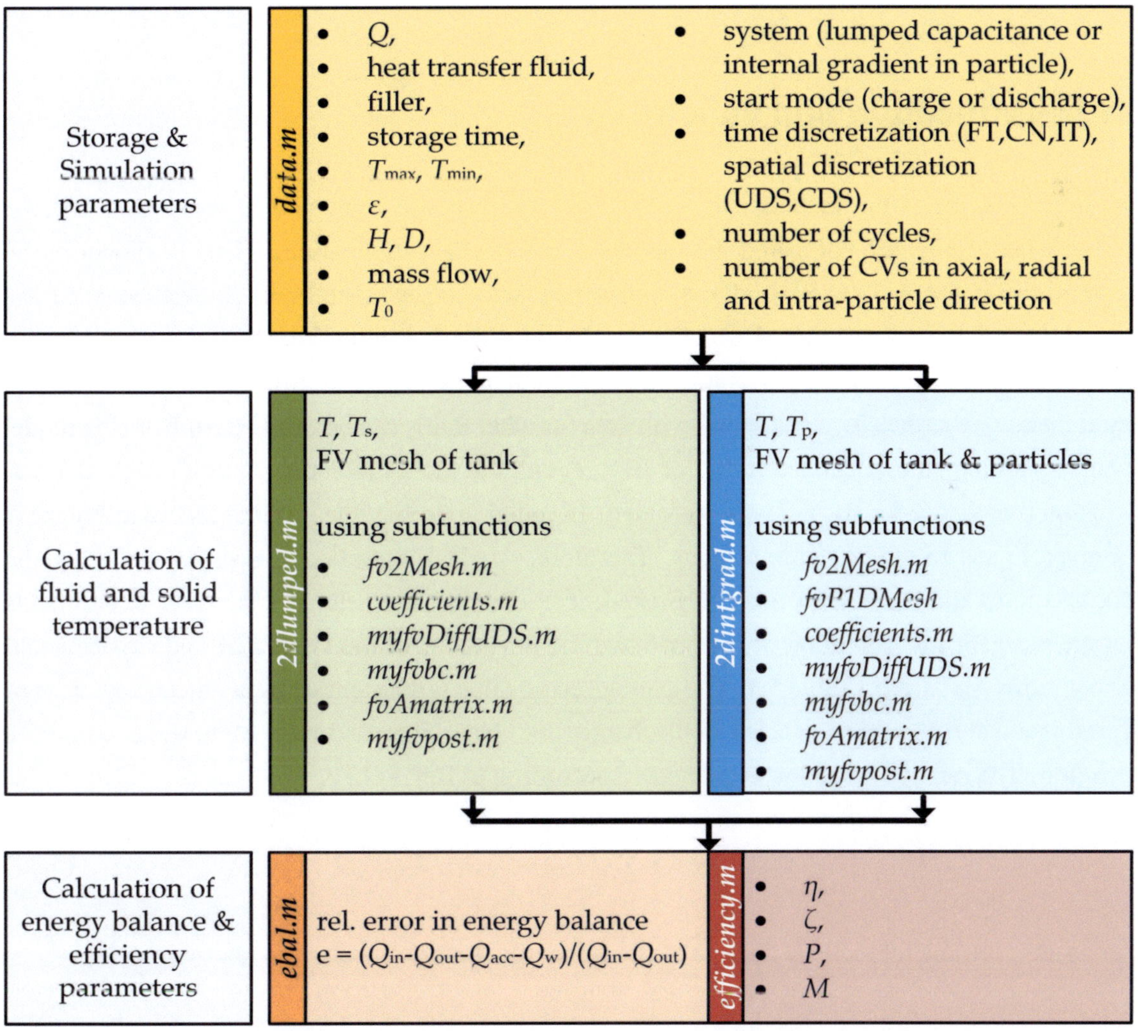

Figure 5.11: Schematic MATLAB program structure

In the sub-function *data.m* all the data are given that characterize the storage system and the parameters needed for the numerical solution. Firstly, the parameters defining the storage system are specified, such as the capacity, heat transfer fluid, filler material and diameter, storage time, maximum and minimum temperatures, porosity, tank dimensions, mass flow and ambient temperature.

Secondly, the system is selected, meaning whether a lumped capacitance can be assumed (Bi < 1) or the intra-particle diffusion needs to be included. The starting mode is defined, which decides the initial condition of the first cycle, meaning whether it is fully charged or discharged at the beginning. Further, the time discretization can be selected and the number of cycles, nodes and time steps is given. Depending on which system is selected, the equations are solved either in the sub-function *2dlumped.m* (lumped capacitance) or in *2dintgrad.m* (including internal gradient in the particles). The main output of these sub-functions is the temperature distributions of the fluid and solid. For the calculation of the temperature distribution, further sub-sub-functions are needed, which are explained in the Appendix (Section A.5.6). Thereafter, the energy balance is checked in the *ebal.m* sub-function and the efficiency parameters (further explained in Section 6.1) are calculated in the *efficiency.m* sub-function.

5.1.5 Verification and validation

The correct setup of the models and their implementation are evaluated in two steps. Firstly, the individual terms are compared with analytical solutions (verification). This is shown in the Appendix (Section A.5.7). Secondly, a comparison with experimental data is performed to test the validity of the overall model and the assumptions made (validation). This is presented in the following. There are no experimental data available in the literature for a packed bed with any liquid metal. Nevertheless, a validation with data for other fluids can be conducted. Both the model assuming a lumped capacitance (only for Bi < 1) and the model including intra-particle thermal diffusion (necessary for Bi $\gg$ 1) are validated. In radial direction, the temperature distribution is not given in any of the literature sources. Therefore, an ideal insulation is assumed and thus, the heat loss term and radial effects are neglected. The storage parameters of the three experiments that are used for the validation of the code and the deviations of the simulated and experimental temperatures are listed in Table 5.4. The average error value is determined as the mean deviation of experimental and numerical data for all discharge time steps. The maximum value is the maximum deviation. The relative deviation is calculated according to Ref. [113]:

$$\text{rMSE} = 1/n\sum((T_{\text{exp}} - T_{\text{mod}})/T_{\text{exp}})^2 \tag{5.32}$$

The figure given in Table 5.4 is the one of the discharge time step with the maximum rMSE.

The first validation has been conducted with experimental data of a pilot-scale 2.3 MWh_{th} molten salt storage with a quartzite/sand packed bed tested at Sandia National Laboratories [39]. These data are frequently used in the literature for validation. The thermo-physical properties of solar salt and quartzite are taken from Ref. [87] and assumed constant in the considered temperature range. Figure 5.12 shows the experimental data (*symbols*) and simulated results (*dashed lines*). The experimental data at t = 0 h are interpolated and taken as initial condition for the simulation. No calibration of model parameters or any modification of the experimental data has been conducted.

Table 5.4: Storage parameters for molten salt and thermal oil packed-bed experiments (taken from Refs. [39, 87, 149]) and the calculated deviations of the numerically calculated values from the experimental data, *relative Mean Square Error, $^+$with $c_{ps} = 1237$ J/(kgK)

Parameter	S.N.L.	Solar One	Promes-CNRS
Capacity	2.3 MWh_{th}	170 MWh_{th}	8.3 kWh_{th}
Fluid	Solar salt	Caloria HT 43	Rapeseed oil
Filler	Quartzite/sand	Granite/sand	Quartzite rock
Discharge time	2 h	8 h	3.5 h
Tank height/diameter	6.1 m/3 m	12 m/18.2 m	1.8 m/0.4 m
Porosity	0.22	0.22	0.39
Particle diameter	0.015 m	0.0046 m	0.012 m
Fluid mass flow	7.0 kg/s	23 kg/s	0.019 kg/s
Upper temperature	396 °C	296 °C	210 °C
Lower temperature	289 °C	179 °C	160 °C
$Bi \approx \lambda_f/\lambda_s$	0.2	0.04	0.04
Average error (K)	3.74	2.71	2.62 (5.00$^+$)
Maximum error (K)	16.46	20.94	10.83 (26.23$^+$)
rMSE* (10^{-4})	0.9	1.6	1.1 (5.9$^+$)

The calculated temperatures differ from the measured temperatures 3.74 K in average and 16.46 K maximum.

Experimental data from the 170 MW_{th} thermal oil storage system with granite rocks/sand of the test plant Solar One, which was in operation for five years in the 1980s [18], are used for the second validation. Figure 5.13 shows the experimental (*symbols*) and simulated data (*dashed lines*). The thermo-physical properties of Caloria HT oil and granite/sand are taken from Ref. [149]. Again, the experimental data at the initial state are interpolated and taken as initial condition for the simulation of the discharge process. The deviation of the calculated to the experimental temperatures is 2.71 K in average and 20.94 K maximum.

A further validation is performed using more recent data from a lab-scale experiment at Promes-CNRS with rapeseed oil as heat transfer fluid and quartzite rock as filler material [113]. The thermo-physical properties of fluid and filler are taken from Ref. [113]. The storage tank consists of a 1.8-m-high packed bed and a fluid buffer of 0.1 m at the top and bottom of the tank. Furthermore, the authors include the tank wall capacity in the total storage capacity (8.3 kWh_{th}). The simulation is performed using only the bed capacity of 6.4 kWh_{th}.

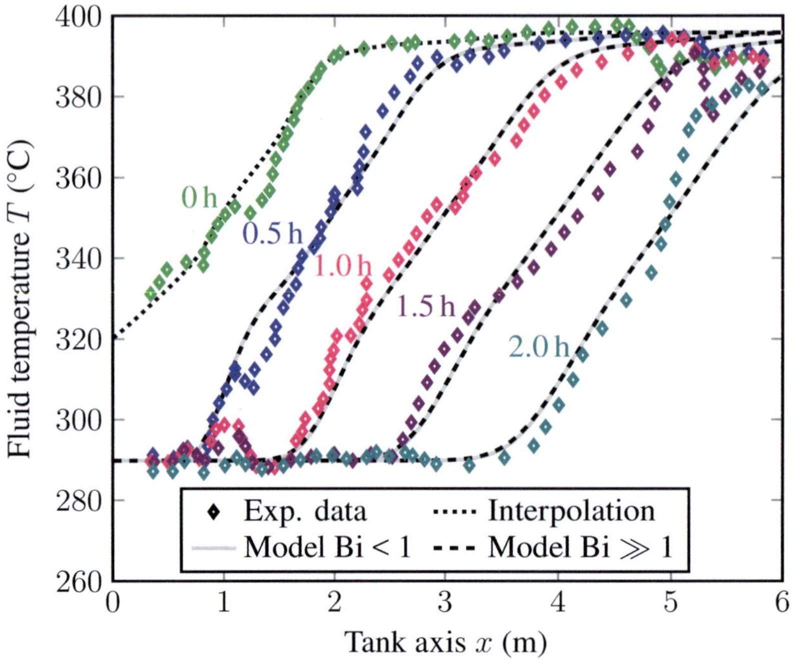

Figure 5.12: Validation of simulation results with experimental data from Sandia Laboratories experiment (taken from Ref. [39])

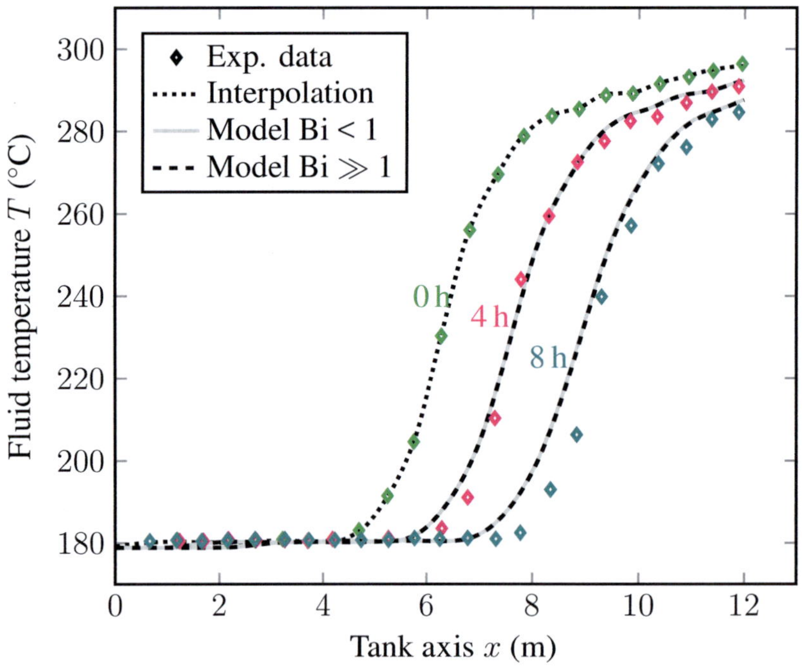

Figure 5.13: Validation of simulation results with experimental data from Solar One (taken from Ref. [149])

The simulated data are in good accordance with the experimental data for the first half of the discharge time, but increasingly differ from the experimental data, as can be seen in Figure 5.14. The deviations are in the same range as the previous cases (Table 5.4). As the tank wall is 23% of the total capacity for this lab-scale experiment (1.9 kWh_{th}) and also releases heat to the fluid during

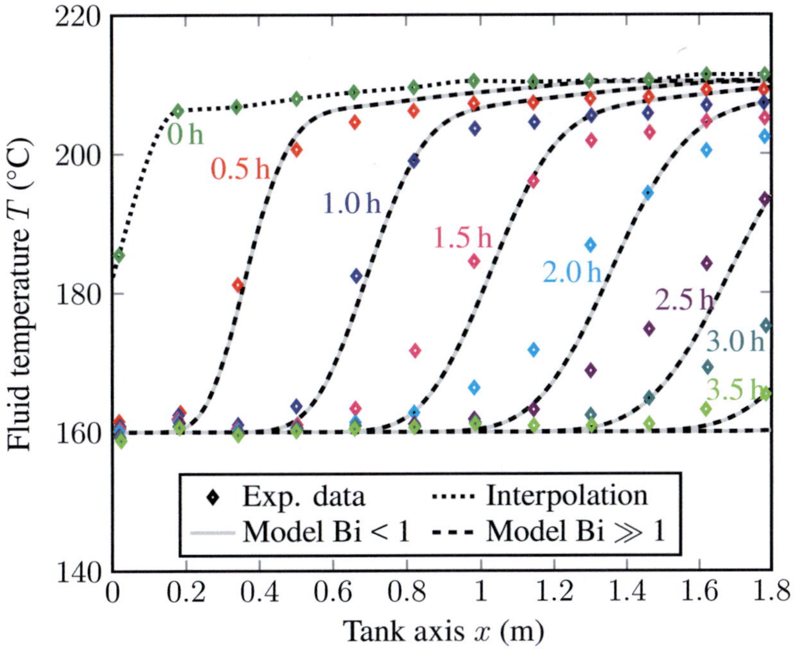

Figure 5.14: Validation of simulation results with experimental data from an experiment of Promes-CNRS (taken from Ref. [113]) with a storage capacity of 6.4 kWh_{th}

discharge, a further simulation is conducted, in which the wall capacity is artificially included in the solid particle capacity. The results are shown in the Appendix (Section A.5.8). They show that now the simulation overestimates the experimental data. It can be concluded that for a lab-scale storage tank, the tank wall capacity needs to be solved for in the simulation, as done by Hoffmann et al. [113].

For both molten salt and thermal oil, no differences resulting from the two different models can be observed in the results; hence, a lumped capacitance is indeed sufficient for those fluids. The meshes used can be found in the Appendix (Section A.7). The average deviations are $\approx 5\,K$ at the most and the relative mean square error is 5.9 10^{-4} at the most. These figures are similar to those obtained by Hoffmann et al. [149] comparing a 1D-2P model without heat losses with their experiments.

In conclusion, all three validation cases show a good agreement of numerical and experimental values, from lab-scale to pilot-scale, even without considering heat losses in the simulation.

5.2 Standby

In addition to the cycling behaviour, the performance during standby behaviour plays an important role (Figure 5.15). Especially for liquid metals, a fast degradation of the thermocline zone is expected, owing to their high thermal conductivity.

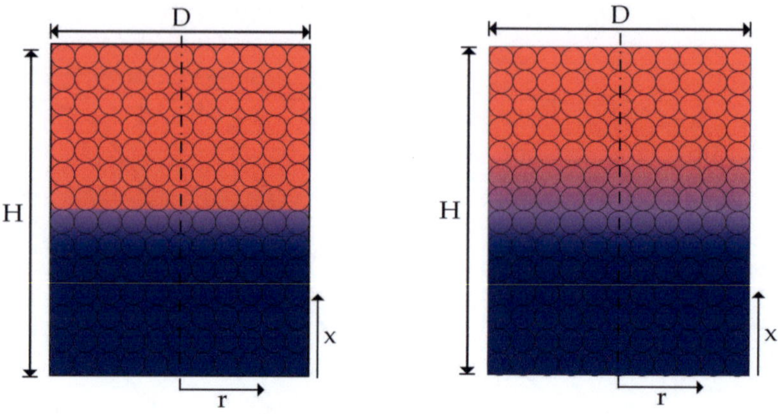

Figure 5.15: Principle scheme of adiabatic tank; left: with an ideal thermocline in half-charged tank (bottom half at lower temperature, upper half at upper temperature); right: with a degradation of the thermocline region after a standby period

In the following sections, the governing energy equations for standby (Section 5.2.1), a comparison of a two-phase and a one-phase model (Section 5.2.2), a comparison of the two-dimensional and one-dimensional model (Section 5.2.3) and the validation (Section 5.2.4) are presented.

5.2.1 Governing equations

The calculation of the thermocline degradation can be conducted similarly to the method for cycling with a two-phase (heterogeneous) or with a one-phase (homogeneous) model. The latter is done in the literature, as shown in Table 5.5. In this work, both approaches are presented and compared. In general, there are far fewer works on standby than on the cycling behaviour of thermocline storage systems, to the best of the author's knowledge.

Table 5.5: Models used in the literature to calculate thermocline degradation during standby, P = phase, D = dimension

Model	General simplification	Investigations	Ref.
1P-1D Analytical	• fluid and solid considered as one phase • radially uniform temperature • thermal losses neglected	• development of analytical equation for cycling and stand by • comparison of molten salt and molten salt/rocks storage	[164]
1P-2D	• fluid and solid considered as one phase • radially uniform temperature • mixed thermal capacity • mixed effective thermal conductivity	• comparison of influence of wall with and without insulation during stand by	[89]

Two-phase (heterogeneous) model. The temperature distribution can be calculated in a heterogeneous model as previously described in Section 5.1.1. The following terms are adjusted:

- No fluid flow ($u = 0$)
- No convective, but only conductive heat transfer (Nu $= 2$) considered in h_v (as already done for liquid metals for cycling)

The energy equations for the fluid phase in the two-phase two-dimensional model including intra-particle thermal diffusion are then reduced to:

$$\varepsilon \rho_f c_{pf} \frac{\partial T}{\partial t} = k_{fx} \frac{\partial^2 T}{\partial x^2} + k_{fr} \left(\frac{\partial^2 T}{\partial r^2} + \frac{1}{r} \frac{\partial T}{\partial r} \right) - h_v (T - T_p|_{y=\frac{d}{2}}) \tag{5.33}$$

The intra-particle energy equation and the boundary conditions remain unchanged are as explained for the cyclic mode (Section 5.1.1). For the lumped capacitance model (Bi < 1) the energy equation for the fluid changes to:

$$\varepsilon \rho_f c_{pf} \frac{\partial T}{\partial t} = k_{fx} \frac{\partial^2 T}{\partial x^2} + k_{fr} \left(\frac{\partial^2 T}{\partial r^2} + \frac{1}{r} \frac{\partial T}{\partial r} \right) - h_v (T - T_s) \tag{5.34}$$

One-phase (homogeneous) model. The homogeneous model, as used in the literature, considers fluid and solid as one phase with mixed properties. Thus, there is no consideration of heat transfer between fluid and solid, but only of thermal conduction in this "mixed" phase.

$$(\rho_f c_{pf})_{mix} \frac{\partial T_{bed}}{\partial t} = k_{mix} \frac{\partial^2 T_{bed}}{\partial x^2} + k_{mix} \left(\frac{\partial^2 T_{bed}}{\partial r^2} + \frac{1}{r} \frac{\partial T_{bed}}{\partial r} \right) - h_{w,v} (T_{bed} - T_0) \tag{5.35}$$

The VDI Wärmeatlas (Chapter D6) [135] gives an overview of different models for the effective heat conduction coefficient in packed beds with stagnant media. There are three different approaches to modelling the mixed thermal conductivity in a packed bed (Figure 5.16).

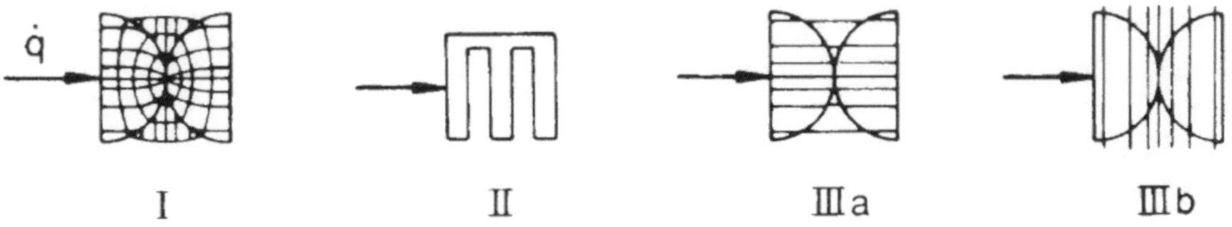

Figure 5.16: Type I: analytical solution; Type II: combination of resistances; Type IIIa: unit cell with parallel heat flow lines; Type IIIb: unit cell with parallel isothermal lines [135]

Firstly, an analytical solution of Maxwell can be used (Type I), which is suitable only if the particles do not affect each other, e.g. in diluted suspensions, good results can be obtained only for $k_p = \lambda_s / \lambda_f < 20$. Secondly, serial and parallel combinations of fluid and solid heat resistances are proposed (Type II). The Krischer model combines both limit cases in a series connection with a weighting factor a (Table 5.6). For packed beds, the factor $a = 0.2$ is recommended in Ref. [135].

With the serial connection the minimum value is calculated; the parallel connection gives the maximum one, while the Krischer model leads to values in between [135]. Thirdly, Type III is a compromise of Type I, which is complex but close to reality, and Type II, which is simple but a rough estimation. This model is exact only for $\lambda_s = \lambda_f$. Especially for liquid metals and the filler material considered in this work, this is by far not the case. Nevertheless, Type IIIa, which is recommended for practical use, is included in this study for comparative reasons. It describes the effective conductivity of a unit cell with parallel flow lines and was developed by Zehner, Bauer and Schlünder (ZBS). The equations for the different models are given in Table 5.6.

Table 5.6: Definitions of effective conductivity k_{mix} used in the literature with $k_p = \lambda_s/\lambda_f$, taken from Ref. [135]

Type	$k_{mix} =$	
I	$k_{Maxwell} = \frac{1+2\varphi}{1-\varphi}\lambda_f$ with $\varphi = \frac{(1-\varepsilon)(k_p-1)}{k_p+2}$	(5.36)
II	$k_{parallel} = k_{max} = (\varepsilon + (1-\varepsilon)k_p)\lambda_f$	(5.37)
II	$k_{serial} = k_{min} = \left(\varepsilon + \frac{1-\varepsilon}{k_p}\right)^{-1}\lambda_f$	(5.38)
II	$k_{Krischer} = \frac{1}{a/k_{serial}+(1-a)/k_{parallel}}$ with $a = 0.2$	(5.39)
IIIa	$k_{ZBS} = (1-\sqrt{1-\varepsilon}+\sqrt{1-\varepsilon}k_c)\lambda_f$ with $k_c = f(k_p, \varepsilon)$	(5.40)

In Figure 5.17, the different models for k_{mix} are displayed as a function of k_p. For the reference case of solar salt and quartzite filler, the different models do not lead to significantly differing results. However, for sodium and the range of filler materials considered in this work, the models show large differences. For example, for sodium and quartzite filler, the maximum (Type II: parallel connection) predicted value is $\approx$ 10 times larger than the minimum one (Type II: series connection).

To understand the influence of k_{mix} on the temperature distribution in the tank during standby, the different k_{mix} figures for sodium/quartzite are used to simulate the thermocline degradation during standby. For the comparison of the different k_{mix} models, the heat losses and radial effects are set to zero, thus, Equation 5.35 is reduced to:

$$(\rho_f c_{pf})_{mix} \frac{\partial T_{bed}}{\partial t} = k_{mix} \frac{\partial^2 T_{bed}}{\partial x^2} \tag{5.41}$$

The initial condition is an ideal thermocline with the bottom half of the tank being at minimum temperature and the upper half being at maximum temperature. As boundary conditions, no heat transfer ($\dot{q} = 0$) is assumed at the in- and outlet of the tank. For the mixed thermal capacity $(\rho c_p)_{mix}$ the definition of Xu et al. [89] is used:

$$(\rho c_p)_{mix} = \varepsilon \rho_f c_{pf} + (1-\varepsilon)\rho_s c_{ps} \tag{5.42}$$

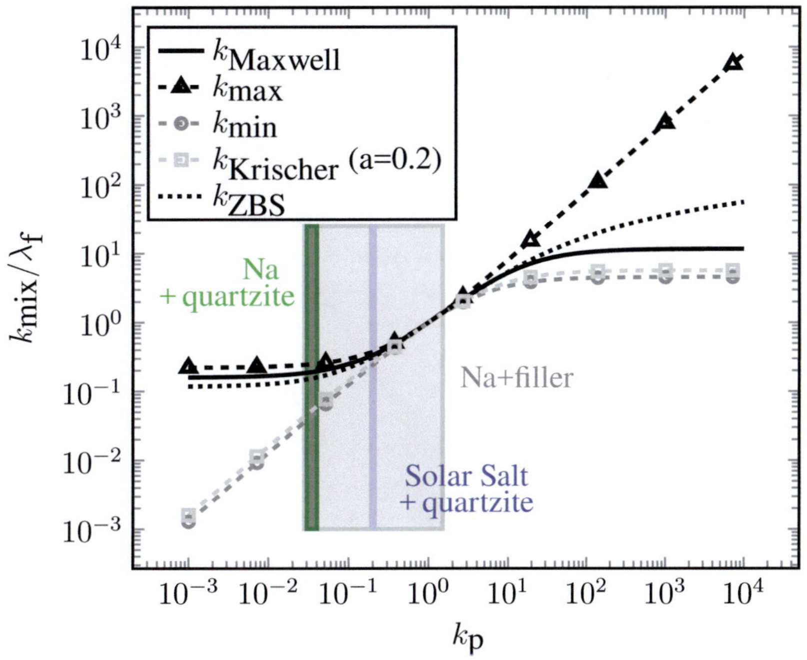

Figure 5.17: Dependency of k_{mix} on $k_p = \lambda_s/\lambda_f$ with porosity of $\varepsilon = 0.22$, the grey area represents the range of k_p for sodium and the filler materials used in this work (list of filler materials in Table A.3)

The storage parameters for the comparison are taken from the Sandia National Laboratories experiment (Table 5.4), with thermo-physical properties of sodium used for the fluid part. In Figure 5.18 the influence of the different models for k_{mix} on the temperature distributions in the sodium/quartzite thermal storage during standby is shown.

With the approach of Maxwell and the model of Zehner/Bauer/Schlünder, k_{mix} are calculated similar to the those with parallel heat resistances and thus, similar temperature distributions are obtained. The Krischer model with $a = 0.2$ and the serial connection of heat resistances lead to significantly less degradation due to a lower k_{mix}. Only approximately a third of the tank is filled by the thermocline region compared with approximately two thirds in the first cases. Therefore, in order to generate the worst case, the model of parallel heat resistances is used in this work.

For solar salt, no such differences can be observed, as the fluid and solid thermal conductivity are in the same range, as shown in the Appendix (Section A.6.1).

5.2.2 Two-phase vs. one-phase model

The results from the two-phase (heterogeneous) and the one-phase (homogeneous) model (with $k_{\max}$) are compared. For this comparison, heat losses are set to zero and radial effects are neglected. The storage tank of the experiment at Sandia National Laboratories [39] is again taken as the reference case with sodium instead of molten salt. The differences in the thermocline degradation after 1 day starting from an ideal thermocline is shown in Figure 5.19.

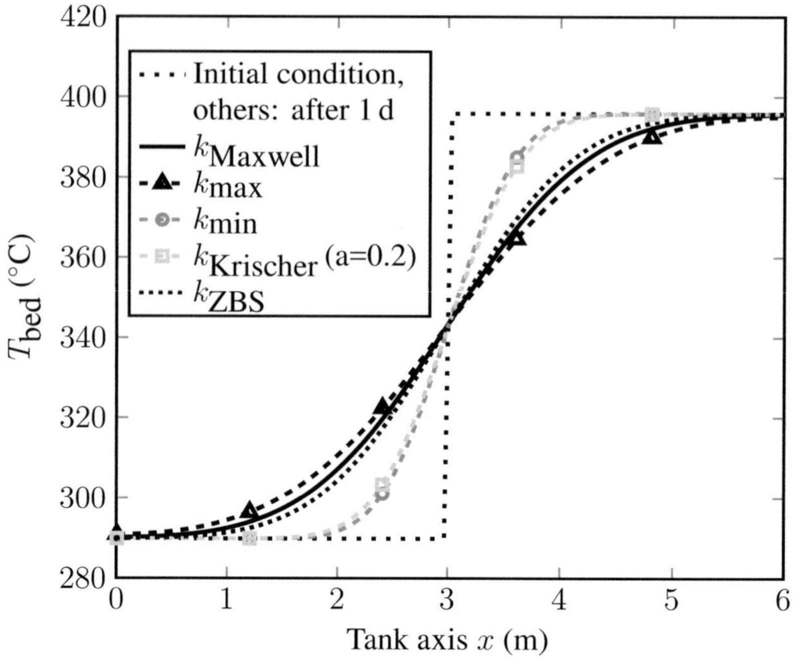

Figure 5.18: Thermocline degradation of a sodium/quartzite packed bed after 1 day of standby resulting from different k_{mix} (at T_{max}); $k_{Maxwell}$ = 14.14 W/(mK), k_{max} = 17.89 W/(mK), k_{min} = 3.17 W/(mK), $k_{Krischer}$ = 3.80 W/(mK), k_{ZBS} = 11.06 W/(mK)

Regarding the heterogenous model, three cases are tested. Firstly, any axial heat conduction from filler to filler is neglected (as only point contact is assumed). Secondly, a (maximum) mixed thermal conductivity in the fluid term is considered (k_{max}). Thirdly, also a mixed thermal capacity $(\rho c_p)_{mix}$ is used in the fluid term. The reason for this approach is that now, in the standby case, the thermal conductivity of the filler can no longer be neglected, as now the thermal processes are not dominated by the convective heat transfer.

The homogeneous and heterogeneous models lead to similar thermocline degradations after 1 day. Using a mixed thermal capacity $(\rho c_p)_{mix}$ leads to the least thermocline degradation, as now the high thermal capacity is included in the mixed term $(\rho c_p)_{mix}$, which is larger than that of sodium in the heterogeneous model. As the thermal conductivity of sodium is already significantly higher than that of the solid ($\approx$ 20 times), the influence of an additionally considered thermal conductivity of the solid is negligible.

In case of solar salt, a lumped capacitance model (Bi < 1) can be used. The results are shown in the Appendix (Section A.6.2). The consideration of axial heat conduction in the solid phase in the heterogeneous model leads to results very similar to those from the homogeneous model with a mixed effective conductivity and mixed thermal capacity. If the axial heat conduction is neglected in the heterogeneous model, this leads to the least thermocline degradation. This results from the relatively high thermal conductivity of the filler material compared with solar salt ($\approx$ 5 times greater), which is not considered. It can be concluded that the homogeneous model can be used for both liquid metals and solar salt to calculate the temperature distributions during standby.

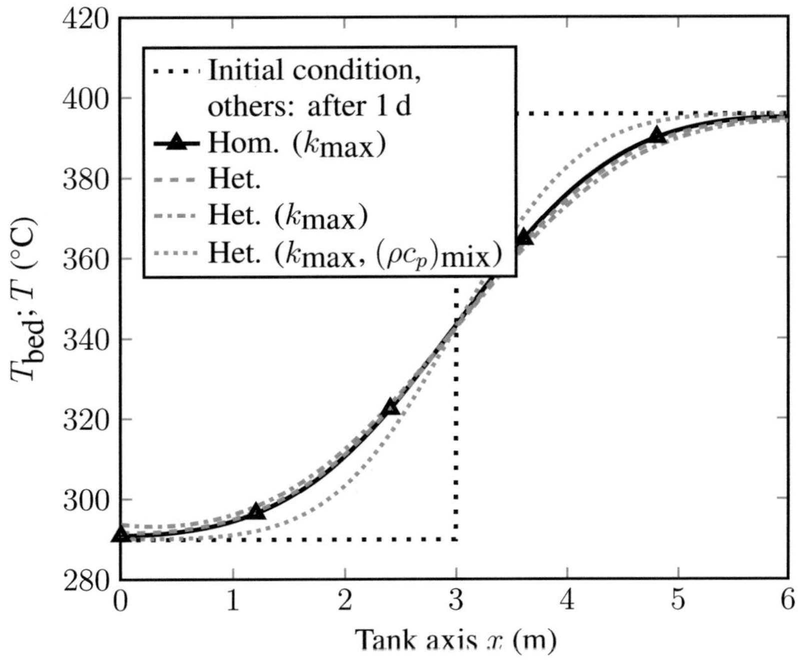

Figure 5.19: Thermocline degradation of a sodium/quartzite packed bed after 1 day of standby resulting from different models, homogeneous with k_{max} and heterogeneous (k_{max}, $(\rho c_p)_{mix}$ in Equation 5.33)

To assess the influence of heat losses on the results of the homogeneous and heterogeneous models, they are compared in the following. Three cases are simulated: Firstly, an ideally insulated tank (h_w = 0 W/(m^{2}K)); secondly, an insulated tank (h_w = 0.24 W/(m^{2}K)); and finally, a non-insulated storage tank (h_w = 9.89 W/(m^{2}K)) (see Section 5.1.3). Again, the storage system of Pacheco et al. [39] is taken as a reference case. The results after 1 day of standby are shown in Figure 5.20. An ideally insulated (adiabatic) and a well-insulated tank show approximately the same result using both the homogeneous and the heterogeneous model, with a temperature difference of maximum $\approx$ 5 K. Without any insulation, the tank cools down towards the ambient temperature (T_0 = 25 °C). The heterogeneous model predicts higher heat losses. This could be caused by the lower thermal diffusivity in the fluid phase in the heterogeneous model ($a = \lambda_f/(\rho c_{pf}) \approx 7 \cdot 10^{-5}$ m^2/s) compared with the mixed thermal diffusivity in the homogeneous model ($a_{mix} = k_{max}/(\rho c_{pf})_{mix} \approx 8 \cdot 10^{-6}$ m^2/s).

This assessment supports two assumptions: Firstly, a well-insulated tank can be calculated as an adiabatic tank. Secondly, a homogeneous model can be used to model the thermocline degradation during standby as well as a heterogeneous model, except for the theoretical case of a non-insulated tank where the heterogeneous model leads to conservative results. The homogeneous model will be used in this work for modelling thermocline degradation processes during standby. For higher temperatures (600–700 °C) the results show the same trends.

For solar salt, the influence of including heat losses in the models is shown in the Appendix (Section A.6.2). The adiabatic and the insulated tank again show similar results. The heterogeneous and the homogeneous models, however, lead to large differences in the thermocline thickness, as

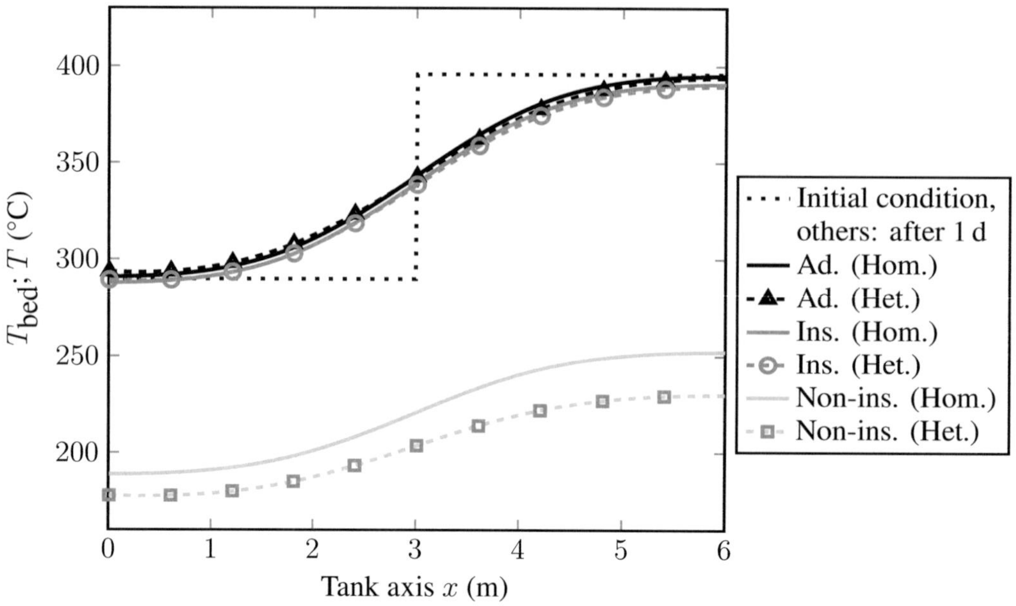

Figure 5.20: Thermocline degradation of a sodium/quartzite packed bed after 1 day of standby resulting from different assumed heat losses, homogeneous (with k_{max}) and heterogeneous (without axial solid conduction)

already observed. The consideration of heat losses leads to the same effect on the homogeneous and the heterogeneous model.

In conclusion, the homogeneous model (with k_{max}) is selected for the simulation of the standby behaviour of a well-insulated tank in the following.

5.2.3 Two-dimensional vs. one-dimensional model

The results from the (homogeneous) two-dimensional (Equation 5.35) and the simplified (homogeneous) one-dimensional model (Equation 5.41 including heat-loss term) are compared in a sodium/quartzite packed bed, based on the storage tank parameters of the Sandia National Laboratories experiment [39]. Again, an ideally insulated, an insulated and a non-insulated storage tank are compared.

Figure 5.21a shows that in the case of an adiabatic system and an insulated system, no differences can be observed along the tank axis at the centre of the tank ($r = 0$). For a non-insulated tank, however, the results along the tank axis differ greatly (up to ≈ 60 K). The temperature distributions in the radial tank axis at the top of the tank are shown in Figure 5.21b, again, only for the non-insulated tank, large differences are observable.

In the 1D model, it is assumed that the heat is lost from the overall volume of the fluid. In the 2D model, by contrast, the heat loss term is considered in the boundary cells. This results in higher heat losses and reduced fluid temperatures when the 1D model is used. In Figure 5.22 the radial effects of the three cases are shown (calculated with the 2D model).

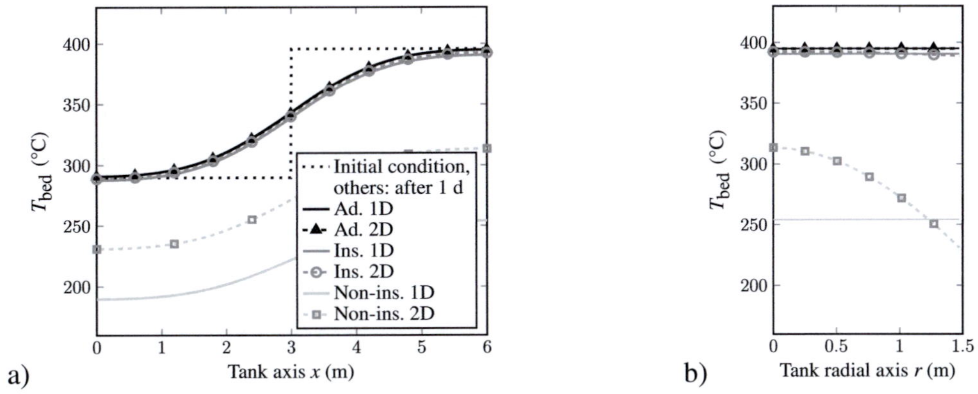

Figure 5.21: a) Axial temperature distributions in a sodium/quartzite packed bed after 1 day of standby resulting from different assumed heat losses in 1D and 2D models (at centre), both homogeneous with k_{max}; b) radial temperature distribution at the top of the tank ($x = H$)

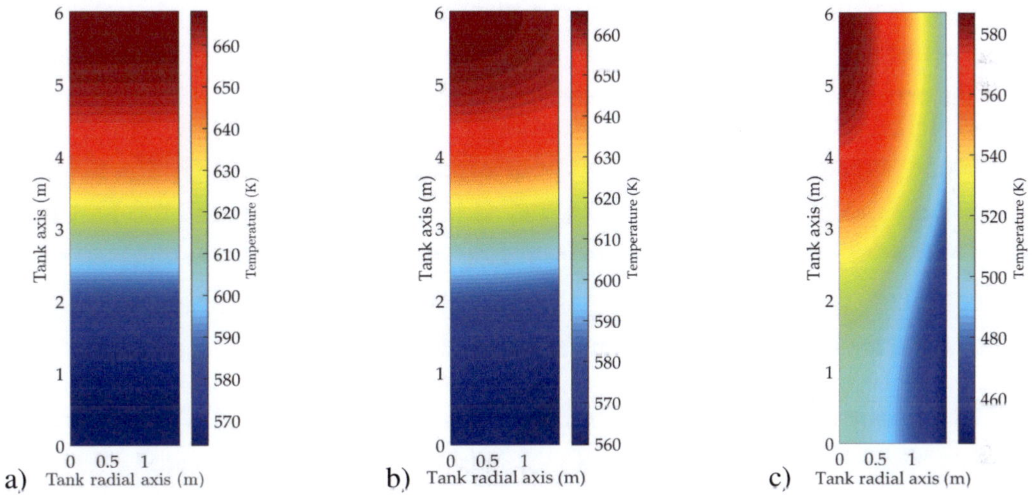

Figure 5.22: Thermocline degradation of a sodium/quartzite packed bed after 1 day of standby resulting from different assumed heat losses, a) adiabatic, b) insulated, c) non-insulated

In the adiabatic case, no radial effects are present and the temperature is uniformly distributed along the radius of the tank, as expected. The insulated tank shows a slight temperature decrease from the centre towards the wall. However, the effect of the thermal diffusion in axial direction is dominant. The temperature distribution in the non-insulated tank after one day is significantly influenced by the wall effects and is now also influencing the axial temperature distribution even at the centre of the tank, as already discussed in Figure 5.21.

From Section 6 onwards, a one-dimensional one-phase model is used to simulate standby processes for ideally and well-insulated tanks. For non-insulated tanks, a two-dimensional one-phase model is used. The initial and boundary conditions are shown in the Appendix (Section A.6.3)

5.2.4 Verification and validation

The verification of the individual terms of the equation with analytical solutions is conducted as shown in the Appendix (Section A.5.7). The validation with experimental data is presented in the following. There is a general lack of data in the literature concerning experiments with thermocline degradation during standby. Only Pacheco et al. [39] show temperature degradation during standby. However, the authors state that "there is a large uncertainly [sic] associated with heat loss measurement because it is based on the change in internal energy of the tank, which cannot be measured very accurately with the instrumentation installed. Also, in this small tank there are heat sinks such as pump penetrations, which were not accounted for in the heat loss model. The penetrations in a larger system will have less affect [sic] on the overall heat loss" [39]. Therefore, the comparison with these data is not expected to be useful for deciding which model (homogeneous/heterogeneous) is more accurate and whether a one-dimensional simulation is sufficient compared with a two-dimensional one, because, due to the relatively small dimensions of the storage, effects of measurement uncertainties and instrumentation acting as heat sinks greatly influence the experimental data, as can be seen in Figure 5.23.

Nevertheless, a comparison of simulated and experimental data is attempted. The homogeneous one-dimensional model is used for this purpose, as it was selected earlier. The heat losses are calculated with a 0.23-m layer of fiberglass insulation (index 1) and a 0.04-m-thick stainless steel wall (index 2) to $h_{\mathrm{w,ins.}} = \left(\frac{s_1}{\lambda_1} + \frac{s_2}{\lambda_2} + \frac{1}{\alpha_{\mathrm{outer}}}\right)^{-1} = 0.42\,\mathrm{W/(m^2K)}$, as described in Section 5.1.3. The heat losses at the top and bottom of the tank are neglected.

Figure 5.23 shows the resulting temperature distributions after 11 h to 41 h. With the estimated heat transfer coefficient $h_{\mathrm{w}} = 0.42\,\mathrm{W/(m^2K)}$ the heat losses are underestimated. With a higher value of $h_{\mathrm{w}} = 1.5\,\mathrm{W/(m^2K)}$, the accordance of simulated and experimental values is improved. All in all, these data are not suitable for validation purposes and do not reflect the heat loss processes in a large tank and further data are not available to the best of the author's knowledge.

5.3 Conclusion

The temperature distributions of the fluid and solid phase during charging and discharging processes can be calculated using a one-dimensional two-phase model including both thermal diffusion in the fluid in radial and axial direction of the tank and heat transfer to the solid particles. For high Biot numbers, as is the case for liquid metals, the intra-particle diffusion needs to be solved as well. For conventional fluids, e.g. solar salt, with low Biot numbers, however, a lumped capacitance can be assumed for the solid phase, and a study showed that it is sufficient to solve an analytical equation without including axial conduction in the solid phase. A comparison of the results with a one-dimensional model, where radial effects are neglected and with a two-dimensional model showed that for both adiabatic and well-insulated tanks, the one-dimensional model leads to similar results. Thus, the simplified one-dimensional model is selected for the parametric studies in the following section. Only for worst-case scenarios (without any insulation), should a two-dimensional model be used. The parameters in the energy equation that is solved for the liquid metal phase ε, u, k_f, h_v and h_w are determined. The parameters are defined as for conventional fluids, except for the heat transfer term between the fluid and the filler, where no Nusselt correlation is available for the heat transfer of low Prandtl number fluids in packed beds. The model has been both verified with analytical solutions and validated with experimental data from molten salt/rocks and thermal oil/rocks experiments.

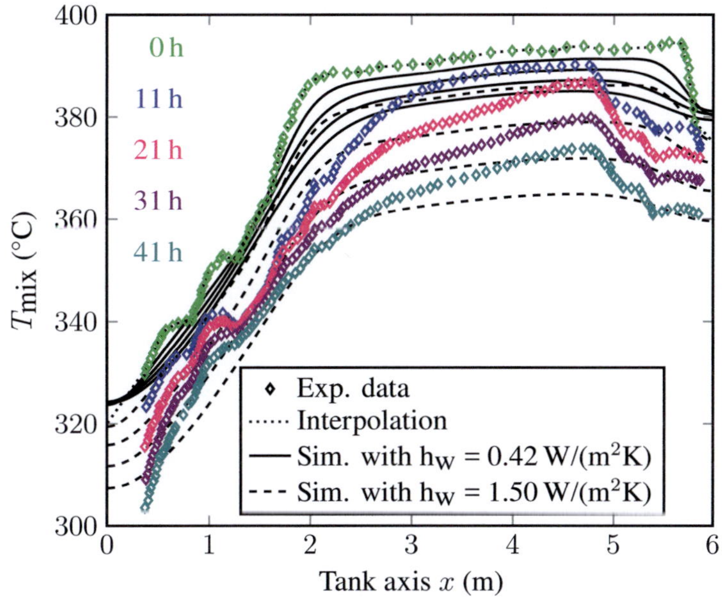

Figure 5.23: Experimental data after 11 h to 41 h of standby of the Sandia National Laboratories Experiment [39] and simulated data with $h_\mathrm{w} = 0.42\,\mathrm{W/(m^2K)}$ and $h_\mathrm{w} = 1.50\,\mathrm{W/(m^2K)}$

During standby, heat conduction is the dominant mechanism. Both a one-phase and two-phase model are discussed to describe the temperature distributions in the tank during standby. For the one-dimensional model, different approaches are compared to describe the "mixed" effective thermal conductivity in the packed-bed storage: an analytical solution of Maxwell, serial and parallel connections of heat resistances, the Krischer model combining the serial and parallel connection and the model of Zehner, Bauer and Schlünder. The parallel connection of heat resistances results in the highest effective thermal conductivities and thus, in the worst case thermocline degradations. A further comparison of the results using a two-dimensional and a one-dimensional model, where radial effects are neglected, leads to the result that the one-dimensional model is again sufficient to model cases with adiabatic and well-insulated storage tanks. For the modelling of heat losses in worst-case scenarios of non-insulated tanks, a two-dimensional model including radial diffusion effects is again recommended. The validation of the individual terms of the standby model was conducted with analytical solutions. The overall model is not validated with experimental data due to a lack of suitable data in the literature.

6 Performance of sodium in a single-tank packed-bed storage

To assess the performance of sodium in a packed-bed storage, firstly, the performance parameters (Section 6.1) and the reference case (Section 6.2) are defined. The results of the parametric study are presented (Section 6.3) and compared with studies for conventional fluids in the literature (Section 6.4). Finally, a comparison of sodium with solar salt and lead-bismuth-eutectic (LBE) for the conventional temperature range (290–565 °C) and chloride and carbonate salts, LBE and lead (500–700 °C) is performed (Section 6.5).

6.1 Evaluation parameters

Two main performance parameters are selected to evaluate the performance of sodium in a packed-bed storage: the discharge efficiency for the cycling mode and the degradation of the thermocline zone for the standby mode. These parameters and the reason for their selection are presented in the following. Further, the methods for determining the required pumping power and the storage medium costs are presented.

6.1.1 Cycling efficiency

In the literature, various definitions of the efficiency during the cycling process of stratified storage systems are proposed [165]. A selection of the most frequently used definitions is shown in Table 6.1.

The definitions can be divided into first-law (energy) and second-law (entropy, exergy) efficiencies. The exergy efficiency gives information not only about the amount of energy extracted but also about the value or quality of this energy. In this work only the energy efficiency parameters are considered.

The energy efficiency definitions can be split into two categories. One is definitions that consider only the outgoing energy flows out of the storage compared with a certain ingoing flow (η and η_{useful}). Among those two, the discharge efficiency η is the parameter most commonly used to evaluate the performance of a packed-bed storage system. The other category is definitions that

Table 6.1: Selection of cycling efficiencies in the literature

Parameter		Definition	Ref.
$\eta = \frac{Q_{\text{dis,out}}}{Q_{\text{chg,in,max}}} = \frac{\int_0^{t_{\text{dis}}} \dot{m}(c_{pf}T_{\text{f,dis,out}}(t) - c_{pf}T_{\text{min}})\,\mathrm{d}t}{\int_0^{t_{\text{chg}}} \dot{m}(c_{pf}T_{\text{max}} - c_{pf}T_{\text{min}})\,\mathrm{d}t}$	(6.1)	Extracted amount of energy during discharge compared with an ideal charging	[88], [87], [93], [117]
$\eta_{\text{useful}} = \frac{Q_{\text{dis,out,useful}}}{Q_{\text{chg,in,max}}} = \frac{\int_0^{t_{\text{dis,cut}}} \dot{m}c_{pf}(T_{\text{f,dis,out(t)}} - T_{\text{min}})\,\mathrm{d}t}{\int_0^{t_{\text{chg}}} \dot{m}(c_{pf}T_{\text{max}} - c_{pf}T_{\text{min}})\,\mathrm{d}t}$	(6.2)	Useful extracted amount of energy ($T_{\text{out}} > T_{\text{max}}$ - 20 K) during discharge compared with an ideal charging	[151], [89]
$\eta_{\text{stored}} = \frac{Q_{\text{stored}}}{Q_{\text{stored,max}}} = \frac{\int_0^{H} (mc_p)_{\text{tot}}(T(x) - T_{\text{min}})\,\mathrm{d}x}{\int_0^{H} (mc_p)_{\text{tot}}(T_{\text{max}} - T_{\text{min}})\,\mathrm{d}x}$	(6.3)	Effective energy stored compared with the maximum energy storable in the tank	[166]
$\Delta S_{\text{gen}} = \Delta S_{\text{store}} + \Delta S_{\text{flow}}$	(6.4)	Entropy as sum of entropy increase in tank and entropy generated by flow	[117]
$\eta_{\text{ex}} = \frac{Ex_{\text{dis,out}}}{Ex_{\text{chg,in,max}}} = \frac{\int_0^{t_{\text{dis}}} \dot{m}c_{pf}\left[(T_{\text{f,dis,out}}(t) - T_0) - T_0 ln\left(\frac{T_{\text{f,dis,out}}}{T_0}\right)\right]\mathrm{d}t}{\int_0^{t_{\text{chg}}} \dot{m}c_{pf}\left[(T_{\text{max}}(t) - T_0) - T_0 ln\left(\frac{T_{\text{max}}}{T_0}\right)\right]\mathrm{d}t}$	(6.5)	Exergy of the bed during discharge compared with exergy of bed during ideal charging	[167]
$\eta_{\text{S}} = \frac{\Delta S_{\text{fully-mixed}} - \Delta S_{\text{real}}}{\Delta S_{\text{fully-mixed}} - \Delta S_{\text{ideal}}}$	(6.6)	Ratio of real and ideal entropy generation with respect to maximum entropy generation	[168]

consider the amount of stored energy content in liquid and solid, e.g. η_{stored}. This efficiency describes a utilization ratio of the storage and can be understood as an attempt to include an economic factor.

In this work, η_{useful} is used to evaluate the storage efficiency. It represents the amount of useful energy above a certain threshold temperature deliverable to the power block compared with an ideal discharge process. A threshold of T_{max} - 20 K is selected, as in Ref. [89]. However, this needs to be adjusted to each process connected. This allows the comparison with a state-of-the-art two-tank system, which would show an ideal discharge behaviour if ideally insulated. The definition of η_{useful} is displayed graphically in Figure 6.1 in comparison with η. The economic factor will be assessed separately as discussed in Section 6.1.4, and not included in a storage capacity during cycling. It has to be noted that η_{useful} is determined only in the stable cycle after a certain number of charge and discharge steps have been performed, each for the complete charge and discharge time.

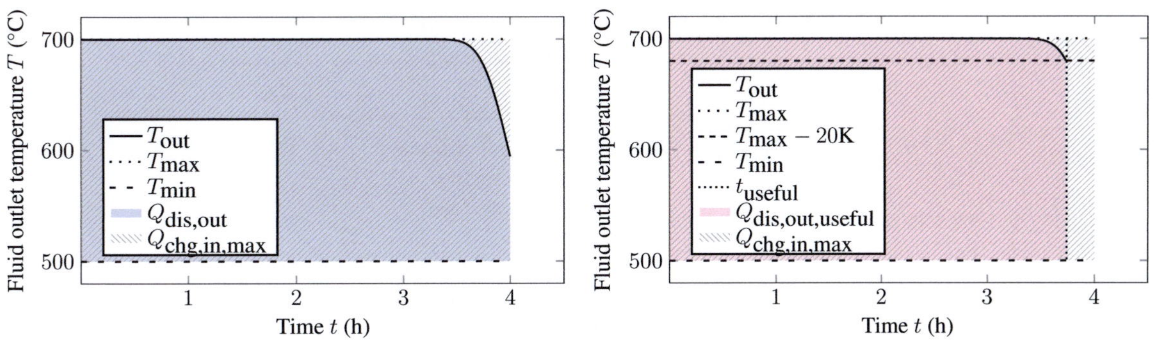

Figure 6.1: Graphical scheme for selected discharge efficiencies; left: $\eta = \frac{Q_{\mathrm{dis,out}}}{Q_{\mathrm{chg,in,max}}}$; right: $\eta_{\mathrm{useful}} = \frac{Q_{\mathrm{dis,out,useful}}}{Q_{\mathrm{chg,in,max}}}$

For an efficient cycling procedure, it can be advisable to vary the charge and discharge time to extract the thermocline region partly or fully in each cycling step, in order to maintain a sharp stratification in the next cycle [115]. These operational strategies are not the subject of this work, as the evaluation of the influence of different storage parameters on the performance of sodium in the selected storage system and the comparison with other fluids are its focus. Therefore, always a full charge and discharge step is simulated.

6.1.2 Degradation during standby

In the literature several figures are proposed to evaluate the standby behaviour of a packed-bed storage system (Table 6.2). The first figure H_{thcl} describes the share of the tank volume, which is neither at minimum nor at maximum temperature, but in the so-called thermocline region, allowing a certain threshold, e.g. $\Delta T = 5\,\mathrm{K}$ [89]. It is related to the total tank height in this work to allow for the comparison of different fluids in storages with changing tank height and a constant capacity.

$$\zeta = H_{\mathrm{thcl}}/H \tag{6.7}$$

The second figure T_{bed} is a graphical display of the temperature distribution along the tank axis and is used to visualize the degradation of the thermocline thickness for different standby times.

The third figure proposed in the literature is used to assess the stratification during a charging/discharging process, but it can also be used for standby.

In this work, the temperature T_{bed} along the tank axis at certain standby times (Figure 6.2 left) and the resulting height of the thermocline zone relative to the tank height ζ (Figure 6.2 right) are used to visualize the thermocline degradation during standby and allow for comparisons of different heat transfer fluids.

Table 6.2: Selection of standby indices in the literature

Parameter		Meaning	Ref.
$H_{\text{thcl}} = \Delta x_{\text{thcl}}(t_{\text{stby}})$	(6.8)	Part of tank volume occupied by the thermocline zone ($T_{\min} + 5 < T < T_{\max} + 5$)	[89]
$T_{\text{bed}} = T_{\text{bed}}(x, t_{\text{stby}})$	(6.9)	Temperature distribution along tank axis after certain standby period	[89]
$\eta_{\text{S}} = \frac{\Delta S_{\text{fully-mixed}} - \Delta S_{\text{real}}}{\Delta S_{\text{fully-mixed}} - \Delta S_{\text{ideal}}}$	(6.10)	Entropy of real storage between ideal case and fully mixed	[168]

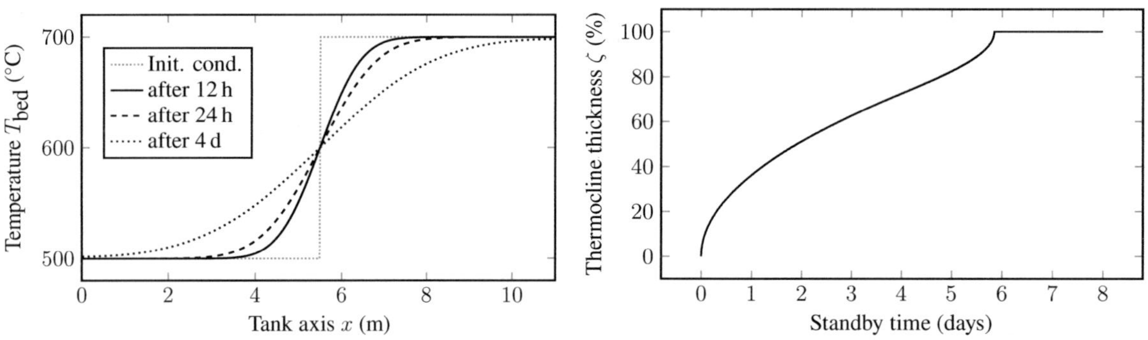

Figure 6.2: Left: Temperature distribution T_{bed}; right: percentage of tank filled by thermocline ζ

In order to obtain the efficiency reduction due to a standby period, the extractable energy from the tank without a standby period is compared with the extracted energy including a typical standby period (8 h). In the latter case, the first half of the discharge step is performed (stable cycle), and after 8 h standby the second half of the discharge step is performed. Then the useful amount of energy η_{useful} is determined as described before.

Figure 6.3 shows the outlet temperature above the threshold of $T_{\max}$ - 20 K without (left) and with a standby period of 8 h after half of the discharge step (right). The "useful" power can be extracted for a shorter time in case of a standby interruption.

6.1.3 Pumping power

The pumping power required for the heat transfer fluid to flow through the packed bed is calculated from the pressure loss and the volume flow of the liquid:

$$P = \dot{V}\Delta p = \frac{\dot{m}}{\rho_{\text{f}}}\Delta p \tag{6.11}$$

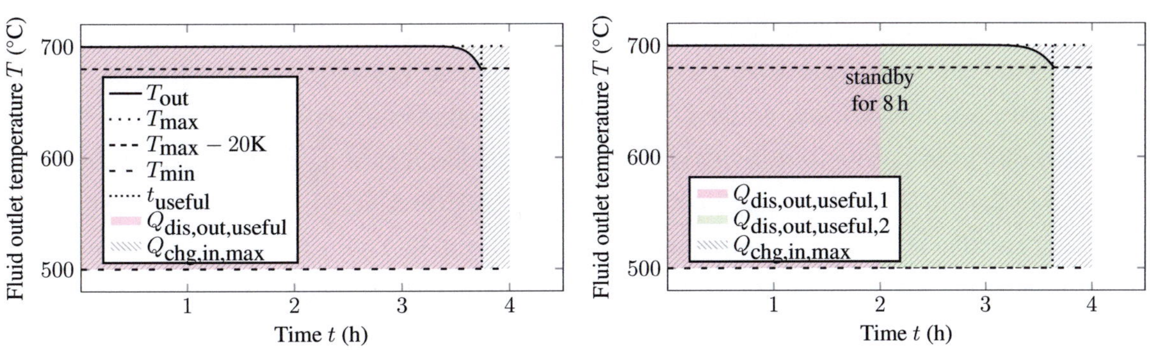

Figure 6.3: Graphical scheme for the extracted energy (shaded area); left: without standby $\eta = \frac{Q_{\text{dis,out}}}{Q_{\text{chg,in,max}}}$; right: including 8-h standby after half of the discharge time $\eta_{\text{useful}} = \frac{Q_{\text{dis,out,useful,1}} + Q_{\text{dis,out,useful,2}}}{Q_{\text{chg,in,max}}}$

The pressure loss through the packed bed results from frictional and inertia losses, which are expressed by linear and a quadratic terms. The most commonly used correlation for this case is the one of Ergun [169] (Equation 6.12). An overview of correlations is given in Refs. [170, 171].

$$\frac{\Delta p}{H} = 150 \frac{(1-\varepsilon)^2}{\varepsilon^3} \frac{\mu_{\text{f}} u_0}{d^2} + 1.75 \frac{(1-\varepsilon)}{\varepsilon^3} \frac{\rho_{\text{f}} u_0^2}{d} \tag{6.12}$$

Ergun and Orning [169] do not explicitly give the validity range of this correlation with respect to the Reynolds number. Allen et al. [170] show that the Ergun equation overpredicts the pressure drop in experiments with $\text{Re}_{\text{p}} > 700$. Stieß [172] gives a wider validity range of $3 < \text{Re}_{\text{p}} < 10^4$. The Reynolds number is for all cases of the single-tank investigations in this section below 700, and thus within the allowable range of Refs. [170, 172]. Only when LBE and lead are used in Section 6.5.2 is the Reynolds number only within the valid range given by Stieß [172]. The constants in the Ergun equation (150 and 1.75) are determined from uniformly sized, spherical particles with small diameters (< 1.5 mm) and with air as HTF, higher velocities and higher porosities as used in the reference case in this study. In packed-bed storage systems, particularly those with sodium, very low bed porosities of $\varepsilon < 0.26$, which are achievable only with non-uniformly sized particles, are desired in order to reduce the required amount of fluid significantly. This is realized by combining small (e.g. sands in mm range) and larger particles (e.g. rocks in cm range). Then, the Sauter diameter can be calculated from the specific surface of the particles s_{v} with Equation 6.13 for spherical and non-porous particles and used in the Ergun equation (6.12) according to the VDI Wärmeatlas [171].

$$d_{\text{Sauter}} = 6/s_{\text{v}} \tag{6.13}$$

Koekemoer et al. [173] present a correlation for the porosity of a non-unifom bed including the difference between largest and smallest particle correlated to experimental data.

Bruch et al. [110] show that only when the pressure loss due to the larger particles is neglected and the sand diameter and the experimentally determined sand porosity instead of the bed porosity in the Ergun equation is used, do the calculated results match the experimental ones well. This is mainly due to the strong sensitivity of the Ergun equation regarding the porosity. A drop of ε from 0.5 to 0.4 (20 %) leads to a pressure drop increase of 300 % [173]. In this work, the pressure losses are calculated with the Ergun equation with an overall bed porosity. The diameter used can be understood as a mean (Sauter) diameter, which also makes it possible to calculate cases for porosities below 0.26. For the comparison of different fluids and to show trends in the parametric study, this approach is sufficient. However, it should be noted that the pressure losses can be underestimated with this assumption and need to be validated in an experiment.

The velocity of the fluid through the packed bed in the Ergun equation is the superficial fluid velocity and is calculated from the mass flow and the diameter of the tank. In Equation 6.14 the tank diameter is already expressed with the D/H ratio, which will be used in the following.

$$u_0 = \frac{4\dot{m}}{\rho_\mathrm{f}\pi(D/H)^2H^2} \tag{6.14}$$

6.1.4 Storage-material cost

A basic economic analysis is included in this study, as the cost reduction of the storage system is one of the main motivations for a packed-bed storage system, as discussed in Section 2.1. The literature review in Section 4.1 regarding thermo-economic studies showed that the cost reduction from a two-tank to a single-tank arrangement is due mainly to the replacement of the often rather expensive liquid storage material with low-cost solid material. According to Pacheco et al. [39], the cost of the storage material accounts for 50% of the overall costs in their study for both two-tank and single-tank arrangements; however, they include only heat exchanger and tank material costs in the overall cost estimation. Libby [99] also includes instrumentation, piping, heat exchangers for indirect storage, engineering costs, contingency, sales tax, etc. with a share of 11–24% for a two-tank and 9–18% for a direct packed-bed storage for the storage material for capacities of 100–3000 MWh.

In the present work, the estimation of the storage-material cost (filler and heat transfer fluid) is taken as a rough but representative measure for the storage cost in this work. However, it has to be kept in mind that, for an overall storage cost estimation, additional costs such as tank material, instrumentation, piping, etc. need to be considered, as well as the total storage capacity, as the costs decrease with increasing storage capacity [99]. Nevertheless, the storage-material cost estimation in this work allows for comparing the costs of fluid-only systems with those of packed-bed systems and for comparing different fluids in a similar storage system with the same capacity and hence, possibly similar additional costs. Possible further costs due to safety measures for the use of sodium are not included.

The storage medium costs M are calculated with Equation 6.15, with C_f being the material cost of the fluid and C_s the cost of the solid material. The costs are related to the storage capacity Q in the tank. The specific material costs of the fluids and solids used in this work are listed in the Appendix (Section A.4).

$$M = \frac{m_f C_f + m_s C_s}{Q} \tag{6.15}$$

6.2 Reference case

A packed-bed storage with a capacity of 40 MWh and a storage time of 4 h is selected as reference case for the parametric study in this work, which is similar to the one investigated by Reddy et al. [88] (see Section 4.1.2). This thermal energy storage system could (ideally) provide a thermal output of 10 MW_{th} for 4 h and an electrical output of 4 MW, assuming an efficiency of 40 % in the power block. The selected storage capacity is in the same range as as the thermal capacity of the sodium tanks (5 MWh, 2 h) used in Almería [174].

6.2.1 Parameters

The following parameters define the reference case. The porosity, D/H, filler material and filler diameter are selected as used in the experiment of Pacheco et al. [39].

- Discharge time: $\Delta t = 4\,\text{h}$
- Storage capacity: $Q = 40\,\text{MWh}_{\text{th}} = 144\,\text{GJ}_{\text{th}}$
- $T_{min} = 500\,°\text{C}$, $T_{max} = 700\,°\text{C}$
- Porosity: $\varepsilon = 0.22$
- Tank diameter-to-height ratio: $D/H = 0.5$
- Heat transfer fluid: liquid sodium
- Filler material: quartzite (SiO_2)
- Filler diameter: $d = 0.015\,\text{m}$

In the parametric study, all these parameters are fixed except for one selected parameter that is varied at a time. Owing to the fixed capacity, this can lead to different tank dimensions, while keeping the tank diameter-to-height ratio constant, e.g. for a varied bed porosity. The storage tank is assumed to be perfectly insulated; therefore, a one-dimensional model is used (see Section 5.1.2).

With all these parameters given, the storage tank height (H), mass flow rate ($\dot{m}$), fluid and solid quantity (m_f and m_s) can be determined.

$$H = \left(\frac{4Q}{\pi (D/H)^2 (\rho_f c_{pf} \varepsilon + \rho_s c_{ps} (1-\varepsilon)) (T_{max} - T_{min})} \right)^{1/3} \tag{6.16}$$

$$m_f = \frac{\varepsilon \rho_f Q}{(\rho_f c_{pf} \varepsilon + \rho_s c_{ps} (1-\varepsilon)) (T_{max} - T_{min})} \tag{6.17}$$

$$m_s = \frac{(1-\varepsilon) \rho_s Q}{(\rho_f c_{pf} \varepsilon + \rho_s c_{ps} (1-\varepsilon)) (T_{max} - T_{min})} \tag{6.18}$$

$$\dot{m} = \frac{Q}{c_{pf} \Delta t \, (T_{max} - T_{min})} \tag{6.19}$$

With physical properties at T_{max}, this results in the parameters given in Table 6.3.

Table 6.3: Storage parameters for reference cases with liquid sodium with physical properties at $T_{max} = 700\,°C$

H (m)	m_f (tons)	m_s (tons)	$\dot{m}$ (kg/s)	u_0 (mm/s)
11.5	53.1	622.2	39.8	1.9

6.2.2 Mesh independency study

A sensitivity analysis for the mesh regarding the number of control volumes (CVs) in the tank axes and in the filler particle was conducted for all configurations (Figure 6.4). In this section, the procedure is illustrated exemplarily for the reference case in the first discharge step, starting from a fully charged tank. Furthermore, no heat losses to the ambient are assumed.

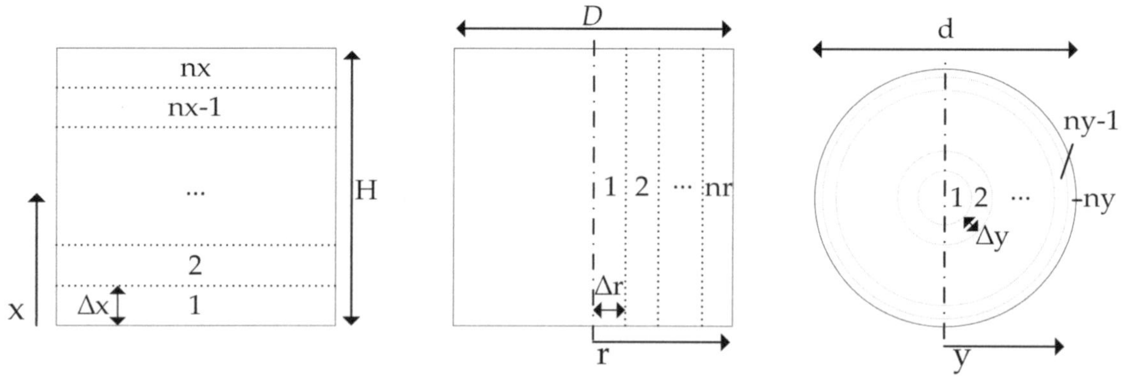

Figure 6.4: Schematic illustration of the mesh in the axial (left) and radial (centre) tank dimension in the tank and in the filler (right)

Firstly, the number of CVs along the tank axis is increased; the corresponding temperature distributions and discharge efficiencies are shown in Figure 6.5. The higher the number of CVs in the tank axis, the higher the slope of the temperature curve. Starting from 2500 CVs ($\hat{=}\ \Delta x = 12\,\text{mm}$), the efficiency changes less than 0.1% from the maximum value obtained and the grid resolution is defined as sufficiently high at this point.

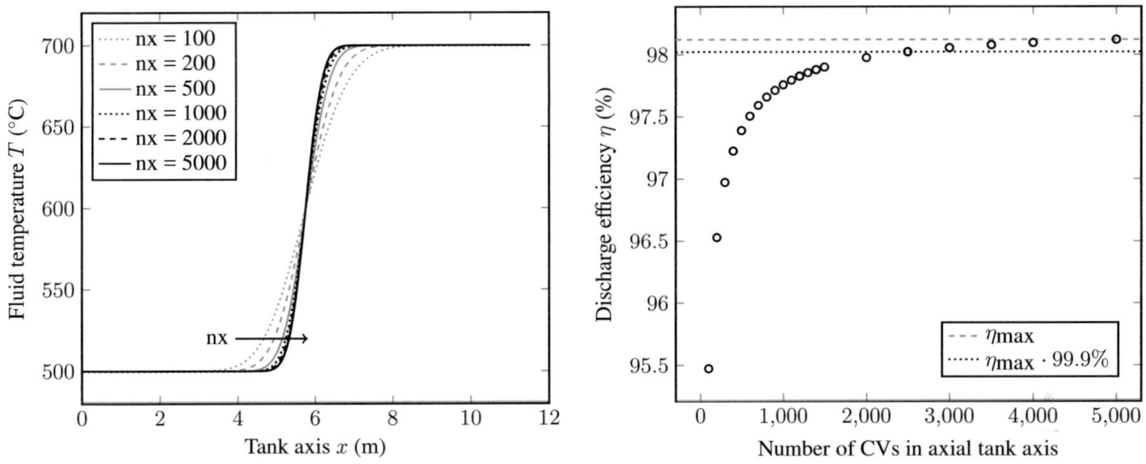

Figure 6.5: Left: Temperature gradient along tank axis at half of the discharge time for increasing number of CVs in the x axis ($\Delta y = 0.8 \ldots 0.0005\,\text{mm}$, $\Delta t = 2.9\,\text{s}$); right: influence of increased CV number in tank axis on discharge efficiency η

In the radial particle axis 70 CVs ($\hat{=}\ \Delta y = 0.8 \ldots 0.0005\,\text{mm}$) are sufficient (Figure 6.6). In the particle, a structured mesh is used with a growth ratio of 0.9 in the direction of the particle surface, where the highest temperature gradients are present. When considering heat losses to the ambient, a mesh study is also performed for the radial direction of the tank in an analogous manner.

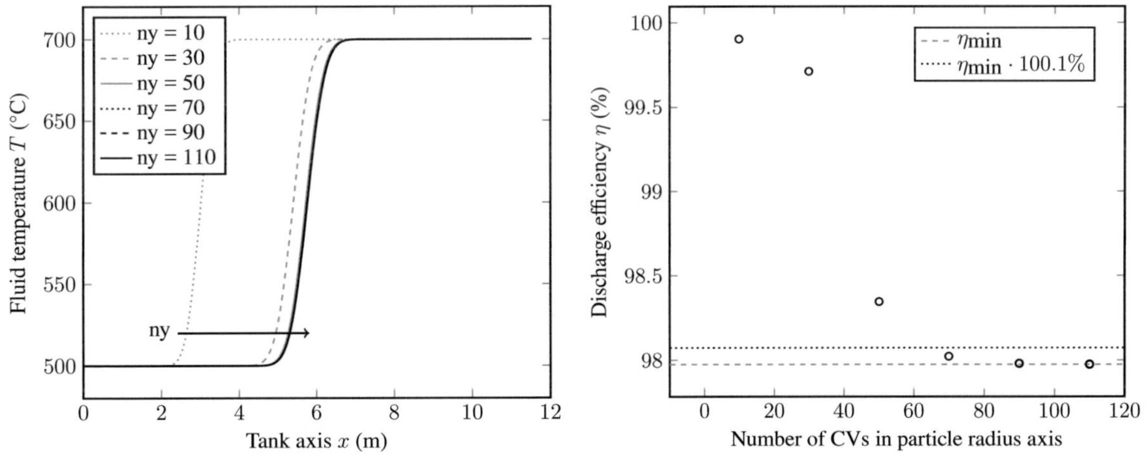

Figure 6.6: Left: Temperature gradient along tank axis at half of the discharge time for increasing number of CVs in the y axis ($\Delta x = 12\,\text{mm}$, $\Delta t = 2.9\,\text{s}$); right: influence of increased CV number in particle radius on discharge efficiency η

The independence of the time step size is investigated by decreasing the time steps from 5 s (3000 steps) to 1.5 s (10000 steps). A time step of $\Delta t = 2.9\,\text{s}$ is sufficient. The variation of the time step size has a negligible influence on the temperature distribution and discharge efficiency in the considered range compared with the CVs in the tank and particles.

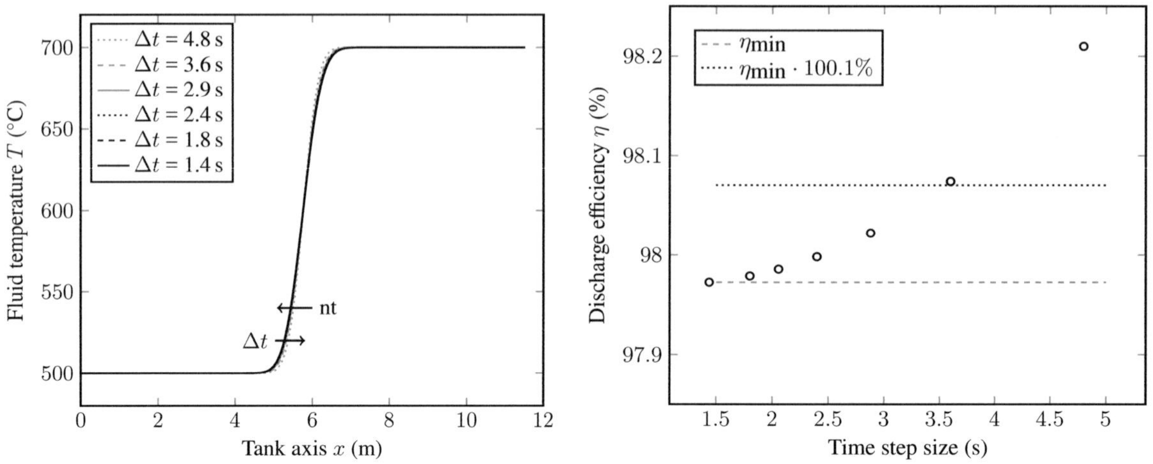

Figure 6.7: Left: Temperature gradient along tank axis at half of the discharge time for different time step sizes ($\Delta x = 12$ mm, $\Delta y = 0.8 \ldots 0.0005$ mm); right: influence of decreased time steps on discharge efficiency η

In the following, the simulations of the reference case are performed with $\Delta x = 12$ mm, $\Delta y = 0.8 \ldots 0.0005$ mm and $\Delta t = 2.9$ s. For all cases of the parametric study, the independence of the solution from the grid is checked again. The mesh parameters for all cases are listed in the Appendix (Section A.7).

6.2.3 Results

Figure 6.8 (left) shows the temperature distributions along the tank axis and the outlet temperature of the fluid during the first discharging step. The steepness of the S-curve of the temperature between minimum to maximum temperature is an indicator of the efficiency of the storage system. The higher the slope of the curve, the longer the fluid can be extracted at maximum temperature and thus, the higher the discharge efficiency η_{useful}. The thermocline region moves through the tank and expands with proceeding discharge time due to the conductive heat transfer along the tank axis. This is illustrated in Figure 6.8 (left) by the grey areas, representing the width of the thermocline region after 1 h and after 3 h of discharging.

The first discharge step differs from the subsequent cycles (Figure 6.8 right), as the first initial condition is of constant maximum temperature and the subsequent initial conditions are the final temperatures of the previous charging step. The fluid temperature in the 4th cycle (discharge) differs less than 0.1% from the previous step and a stable cycle is reached. The resulting discharge efficiencies η_{useful} are presented in Table 6.4.

Figure 6.9 shows the temperature difference inside the filler particles for certain discharge time intervals. An intra-particle temperature difference is obtained only in the thermocline zone. For

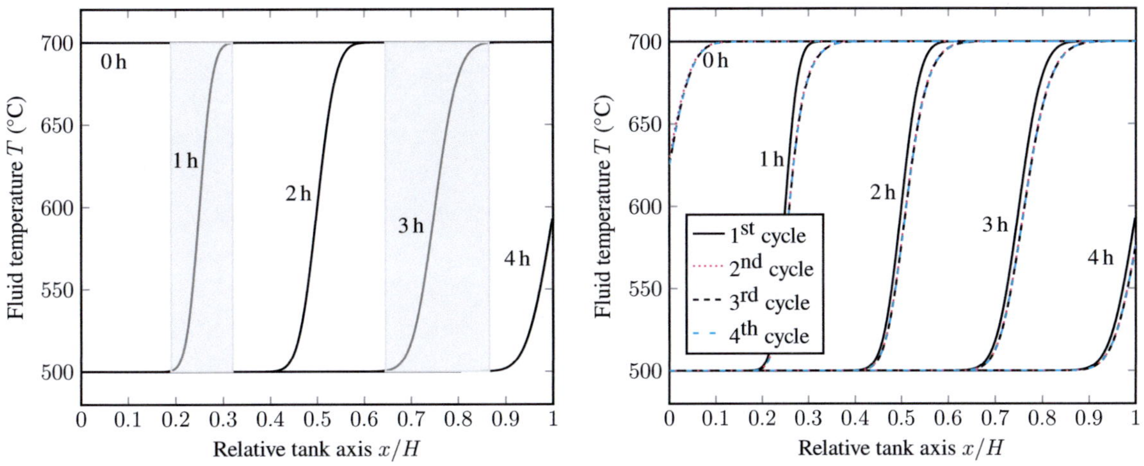

Figure 6.8: Fluid temperature along relative tank axis at 1 h, 2 h, 3 h and 4 h discharge time intervals; left: results for first discharge of completely charged tank; right: results for first 4 discharge cycles

Table 6.4: Discharge efficiencies in first cycles

Cycle no.	1	2	3	4
η_{useful} (%)	93.85	92.07	91.95	91.93

example, after half of the discharge time (2 h) the thermocline region is at the center of the tank at $H/2$ and there is a temperature gradient of $\approx$ 2 K present in the particle. The temperature difference decreases with increasing discharge time owing to thermal diffusion in the tank, leading to a growing thermocline zone.

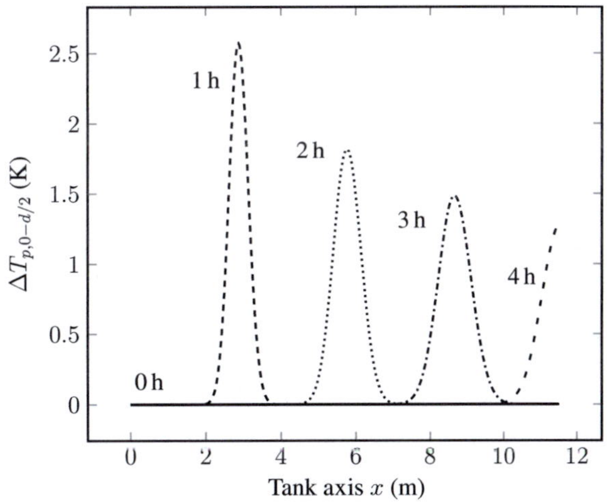

Figure 6.9: Temperature difference between core temperature ($y = 0$) and surface temperature ($y = d/2$) of solid particles at 1 h, 2 h, 3 h and 4 h discharge time (first cycle) along the axis of the tank

In Figure 6.10 the fluid outlet temperatures in a stable cycle with (*solid line*) and without any standby period (*dotted line*) are shown. They directly influence the amount of useful energy that can be extracted (η_{useful}).

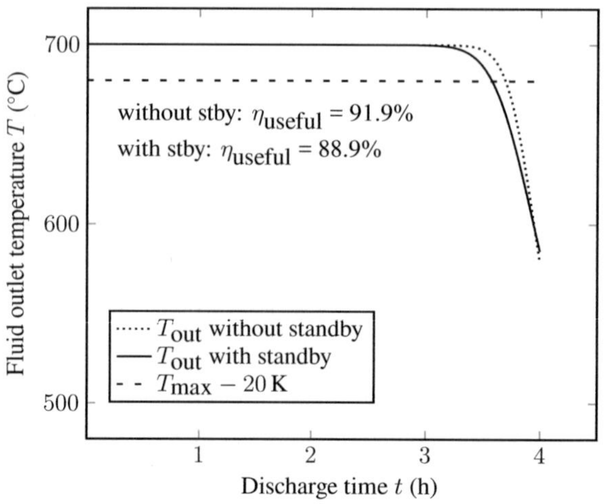

Figure 6.10: Fluid outlet temperature (extracted at top of the tank $x = H$) in a stable cycle without and with standby

In case of an included standby period during discharge, the efficiency is reduced from 91.9% to 88.9% due to thermal conduction in the tank, which leads to a thermocline thickness of 25.7% of the tank height after an 8-h standby period (Table 6.5).

The pumping power is negligible when compared with the electric power that could be used from the thermal energy storage system: 22 W/4 MW = 0.0006%. The Ergun equation that is used for the calculation of the pressure loss is valid for the reference case according to Allen et al. [170] and Stieß [172] owing to the low Reynolds number ($Re_p \approx 114$).

The storage-material costs alone are 11.2 €/kWh. With added piping, etc. it is unlikely that this reference case will fulfill the target of 15 €/kWh mentioned in Section 2.1. The parametric study will show potentials to further reduce the storage-material costs.

Table 6.5: Resulting evaluation parameters in a stable discharge cycle for the reference case parameters

	Ref. case
Discharge eff. η_{useful} (%)	91.9
Thcl. after 8 h standby ζ (%)	25.7
Discharge eff. η_{useful} incl. 8 h standby (%)	88.9
Pump. power P (W)	22.0
Cost M (€/kWh)	11.2

6.2.4 Bidisperse packed beds

For further cost reductions the amount of the relatively expensive heat transfer fluid, and thus low porosities below 0.26 are desired, which can be realized only in bidisperse or multidisperse packed beds. However, the simulations in the literature, and also in this work, are usually done using a mean diameter. Tombrink [175] shows both numerical and experimental results of heat transfer and pressure drop of water in a multidisperse limestone packed bed. This combination is chosen as it behaves like molten salt in a basalt packed-bed storage and is a preliminary work for the TESIS facility of DLR. For the heat transfer, a lumped capacitance model (Bi < 1) is used with differential equations for each solid particle diameter. The results from this approach are compared with results using the Sauter diameter as a mean diameter. They show that the higher the ratio of the diameter of the large to the small particles, the higher the difference in the calculated temperatures and hence, in the discharge efficiencies. Furthermore, the results from using the Sauter diameter and using differential equations for each particle size are compared with experimental results (thermocline gradients) in a bi- and tridisperse packing. Using a mean diameter leads to more than two times larger deviations from the thermocline gradient obtained in the experiment.

In this work, a short study concerning a bidisperse packed bed with sodium is performed. Results from considering two different diameters simultanously are compared with results using a Sauter diameter. The changes in the model, when two different particle sizes are considered, are shown in detail in the Appendix (Section A.5.9). The storage parameters of the reference case are used (Section 6.2), but now with a bidisperse packed bed as used by Pacheco et al. [39] to realize a porosity of $\varepsilon = 0.22$: 30 wt% sand with diameter $d_1 = 1.5$ mm and 70 wt% quartzite rocks with diameter $d_2 = 19$ mm. The corresponding Sauter diameter is $d_{32} = 4.2$ mm (Section A.5.9). In Figure 6.11 the corresponding fluid temperatures along the tank axis are shown. For comparative reasons, the fluid temperatures in a packed bed of 19-mm particles (monodisperse) and 1.5-mm particles (monodisperse) are added.

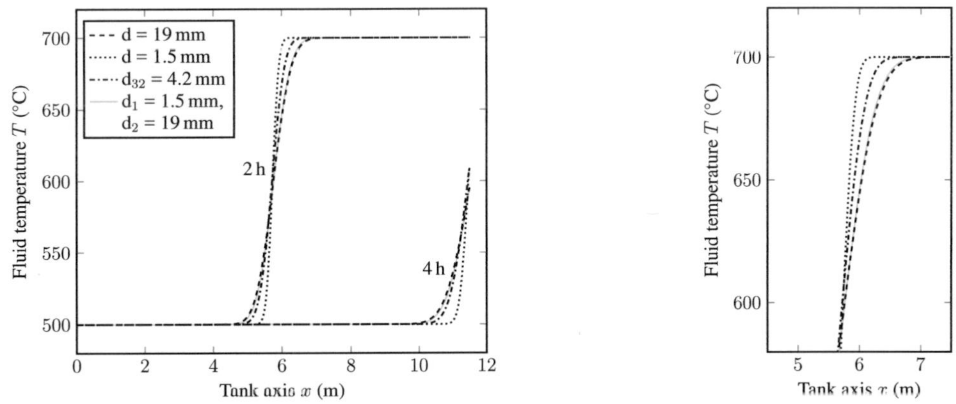

Figure 6.11: Left: fluid temperatures after 2 h and 4 h discharge time (first cycle) with different particle diameters in monodisperse (19 mm, 1.5 mm, 4.2 mm) and a bisdisperse packed bed ($d_1 = 1.5$ mm and $d_2 = 19$ mm); right: zoomed in on fluid temperatures after 2 h

It can be concluded that the fluid temperature with the Sauter diameter differs from the results with two different particle sizes. The heat transfer is limited at the largest particles, which are ≈ 13 times larger than the small particles and thus, the heat resistance inside the particles $\frac{1}{\alpha_s} = \frac{d/2}{\lambda_s}$ is 13 times larger. Therefore, the simulation with the Sauter diameter instead of two different diameters would overestimate the heat transfer between filler and fluid, leading to a smaller thermocline region and thus, an overestimated discharge efficiency η_{useful}. In this case, the latter is 95.7% (d_{32}) compared with 93.5% (d_1,d_2) in the first cycle. This has to be kept in mind when designing a storage system for experimental validation. However, in the parametric study in this work a mean diameter is used for the simulations to show trends and compare different fluids in the selected storage system.

In the Appendix (Section A.5.9) an analogue comparison for solar salt is shown including a comparison with experimental data from Sandia National Laboratories. The same conclusions concerning using a Sauter diameter in the simulation instead of two different filler diameters can be drawn as for sodium.

6.3 Results of parametric study

The literature review in Section 4 indicates that the bed porosity, tank dimensions and the filler size are the parameters with the highest influence on the performance of the packed-bed storage. Therefore, these parameters are investigated in Sections 6.3.1–6.3.3. Parts of the results have already been published in Ref. [176]. The range of the parameters is extended here and additionally, different filler materials, velocities and another temperature range will be investigated in Sections 6.3.4–6.3.6. In Table 6.6 the influence of the varied parameters on the selected performance parameters is presented[4].

Table 6.6: Varied parameters, their range and the affected evaluation parameters (reference case bold)

Parameter	Range	η	ζ	P	M
ε	[0.1,**0.22**,0.4,0.6,0.8,0.95]	✓	✓	✓	✓
D/H	[0.1,0.3,**0.5**,1.0,1.5,2.0]	✓	✓	✓	–
d (mm)	[1,5,**15**,30,50]	✓	✓	✓	–
Filler material	[**SiO_2**,Mg_2AlO_4,Al_2O_3,Steel,Fe]	✓	✓	✓	✓
u_0 (mm/s)	[0.5,1.0,**1.9**,3.5,5.0]	✓	✓	✓	–
T (°C)	[290–565,462.5–737.5,**500–700**]	✓	✓	✓	✓

[4] As a preliminary work, parametric studies with sodium in a packed-bed storage, however using a different reference case, have been performed within the framework of the Master thesis of N. Zancan (2017).

6.3.1 Influence of porosity ε

An increased porosity leads to a larger amount of fluid in the packed-bed system. Due to the smaller thermal capacity (ρc_p) of sodium compared with the solid filler material, this results in larger tank dimensions while keeping the storage capacity constant. The resulting discharge efficiencies, thermocline thickness during standby, pumping power and storage-material costs are shown in Figure 6.12.

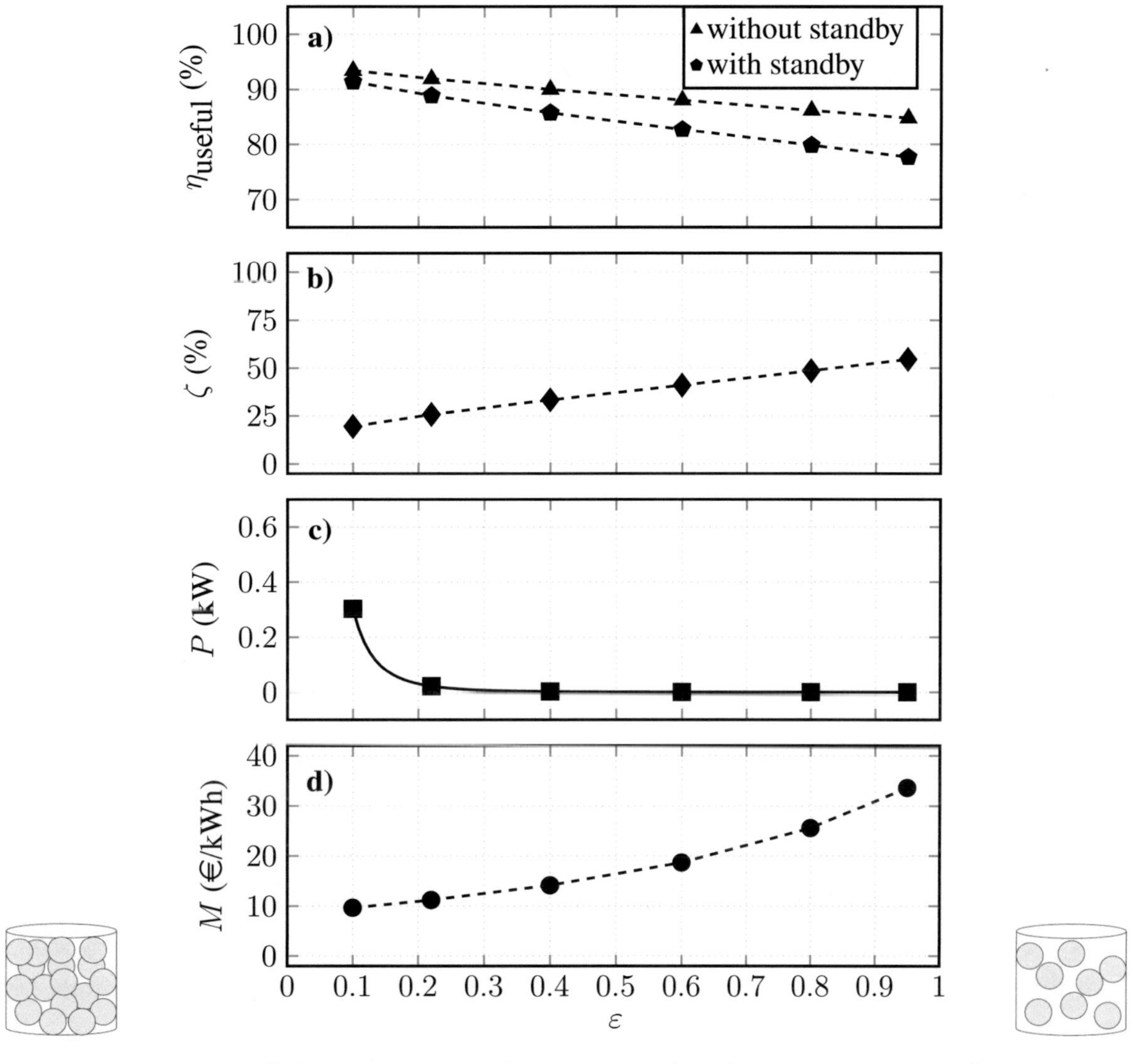

Figure 6.12: a) Discharge efficiency η_{useful} vs. porosity ε; b) thermocline thickness after 8-h standby ζ vs. ε; c) pumping power P vs. ε; d) storage-material cost M vs. ε

Lowest porosities lead to highest discharge efficiencies owing to the large amount of poorly conducting (compared with sodium) solid material. At the same time, even for a porosity of 0.1, the heat transfer between solid and fluid is still sufficient, thanks to the high thermal conductivity of the liquid metal. If an 8-h standby is included after half of the discharge step, the amount of extractable useful energy is reduced, leading to a discharge efficiency of 93.4% (without standby) compared with 91.4% (with standby) for $\varepsilon = 0.1$. The effect of standby on the discharge efficiency is greater for higher porosities. For $\varepsilon = 0.95$ the efficiency is reduced from 84.8% to 77.7%. The thermocline thickness (after 8 h of standby) increases with increasing porosity; from 19.5% (2.1 m) for $\varepsilon = 0.1$ to 54.6% (8.2 m) for $\varepsilon = 0.95$. This is due to a higher share of liquid sodium with its high thermal conductivity, which leads to an overall increased thermal diffusivity a_{mix}.

The pumping power is increasing with low porosities, but for $\varepsilon = 0.1$ it is still < 0.007% of a possible electric output of 4 MW. The storage-material cost is lowest for lowest porosities owing to a large amount of low cost filler material. All in all, low porosities are desirable regarding both efficiency and material costs, while accepting an increased pumping power.

In this parametric study, simulations are performed for porosites below 0.26. These values are achievable only by using a multidisperse packed bed [177], as explained in Section 6.2.4. Nevertheless, the validation with experimental data (Section 5.1.5) with a porosity of 0.22 shows good agreement with using a mean diameter.

6.3.2 Influence of D/H

Varying the parameter D/H while keeping the storage capacity constant leads to changed tank diameters and heights. As the amounts of fluid and solid are constant, the storage-medium costs are not influenced. The resulting discharge efficiencies, thermocline thickness during standby and pumping power are presented in Figure 6.13.

A reduced D/H ratio, meaning small tank diameters and large tank heights, is beneficial with regard to the discharge efficiency. Without standby, a discharge efficiency η_{useful} of 94.4% can be achieved for $D/H = 0.1$. For $D/H = 2.0$ the efficiency is reduced to 83.9%. In tanks with larger tank heights a smaller part of the tank is occupied by the thermocline region, resulting in an improved discharge efficiency. With a standby period of 8 h, the efficiencies are reduced to 93.8% ($D/H = 0.1$) and 75.1% ($D/H = 2.0$). The thermocline thickness after 8 h standby is best for low D/H ratios. For $D/H = 0.1$, 12.4% (4.2 m) of the tank is occupied by the thermocline zone compared with 61.1% for $D/H = 2.0$ (2.8 m). In the Appendix (Section A.6.4) results are shown on whether the tank with the lowest D/H ratio, and thus the largest wall surface, can still be assumed as adiabatic. It is shown that only a 7 K temperature difference is present compared to the results with an adiabat tank. Although best efficiencies are reached with a minimum D/H ratio, this leads to highest pumping powers in consequence of the higher velocity. Furthermore, it should be noted that in case of $D/H = 0.1$, the tank height is $\approx$ 34 m (for $D/H = 0.2$ the tank height is $\approx$ 21 m). This exceeds the state-of-the-art tank height, which is ca. 20 m [178], but is still worth investigating from a theoretical point of view.

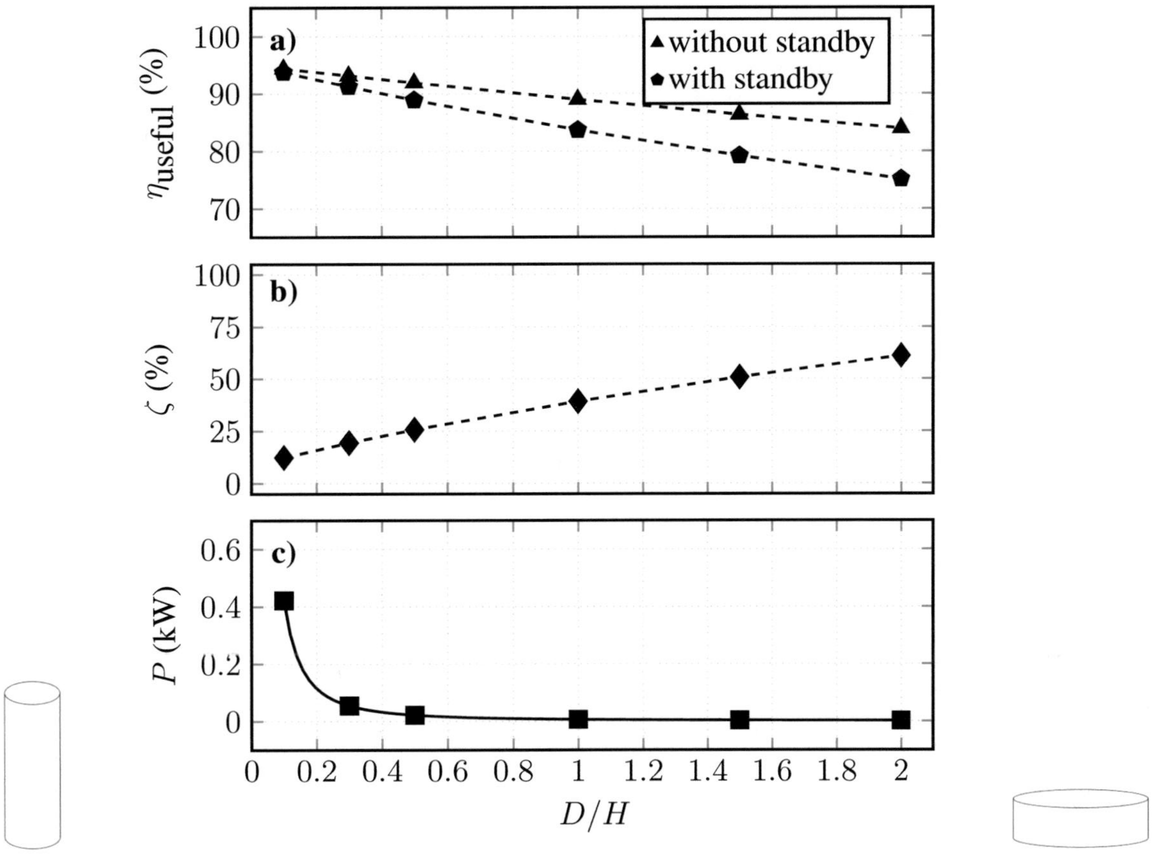

Figure 6.13: a) Discharge efficiency η_{useful} vs. tank diameter-to-height ratio D/H; b) thermocline thickness after 8-h standby ζ vs. D/H; c) pumping power P vs. D/H

6.3.3 Influence of filler diameter d

The variation of the filler particle diameter d has no effect on the tank geometries and the amounts of storage material and therefore no influence on the storage-material costs, when the capacity is kept constant. However, it does influence the discharge efficiencies, the thermocline thickness and the pumping power, as shown in Figure 6.14.

Highest discharge efficiencies are reached for the smallest filler particle diameters. For d = 1 mm the discharge efficiency η_{useful} is 96.7% compared with 83.4% for d = 50 mm. Smaller filler particles lead to both an increased heat transfer coefficient α (Equation A.64) and a larger specific surface s_{v} (Equation A.63) and hence, an increased volumetric heat transfer coefficient: $h_{\text{v,d=1 mm}} = 5.4\,10^8\,\text{W/(m}^3\text{K)}$ and $h_{\text{v,d=50 mm}} = 2.2\,10^5\,\text{W/(m}^3\text{K)}$. Additionally, a smaller filler particle diameter results in a reduced heat transfer resistance inside the particles ($\frac{1}{\alpha_s} = \frac{d/2}{\lambda_s}$), leading to a reduced temperature gradient inside the particles, as shown in Figure 6.15, where the internal temperature difference in the particles after half of a discharging step is presented.

For the reference diameter, the temperature difference is $\approx 2\,\text{K}$ and for the maximum diameter it reaches $\approx 10\,\text{K}$.

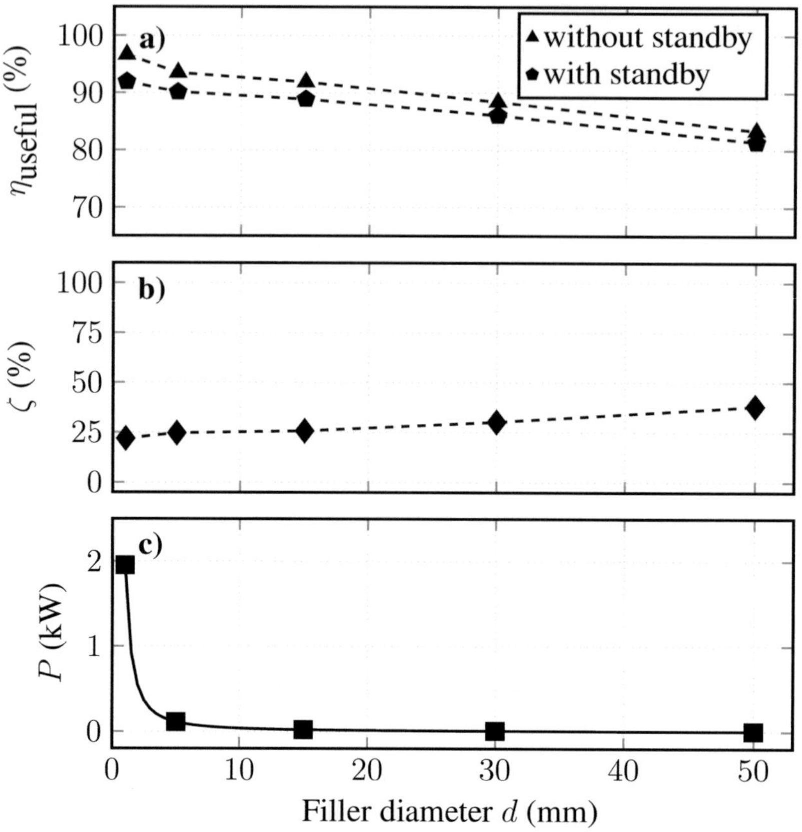

Figure 6.14: a) Discharge efficiency η_{useful} vs. filler particle diameter ratio d; b) thermocline thickness after 8-h standby ζ vs. d; c) pumping power P vs. d

The discharge efficiency is reduced to 92.0% for d = 1 mm and 81.4% for d = 50 mm with an included standby period. The thermocline zone increases to 21.9% for the smallest and to 38.1% for the largest particles in a standby period of 8 h. It should be noted that the one-phase model used for the standby simulation does not include any influence of the filler diameter. Therefore, only the different initial condition of the standby period (after half of a stable discharge step) leads to different results for different filler diameters. A comparison with the results from using the heterogeneous model in the standby simulation for the largest diameter of 50 mm, where the largest difference is expected, shows no significant difference from the two models: The heterogeneous model leads to ζ = 38.2% compared to 38.1% with the homogeneous model.

Again, the pumping power is highest at conditions leading to the highest efficiency but still < 0.05% of an electric output of 4 MW.

6.3.4 Influence of the filler material

Changing the filler material while keeping the overall storage capacity leads to changes of the dimensions of the storage tank owing to the physical properties of the filler material, as shown in

Table 6.7. The physical properties can be found in Table A.3 in the Appendix. The filler materials are numbered from "1" to "5" in increasing order of the thermal diffusivity a_s.

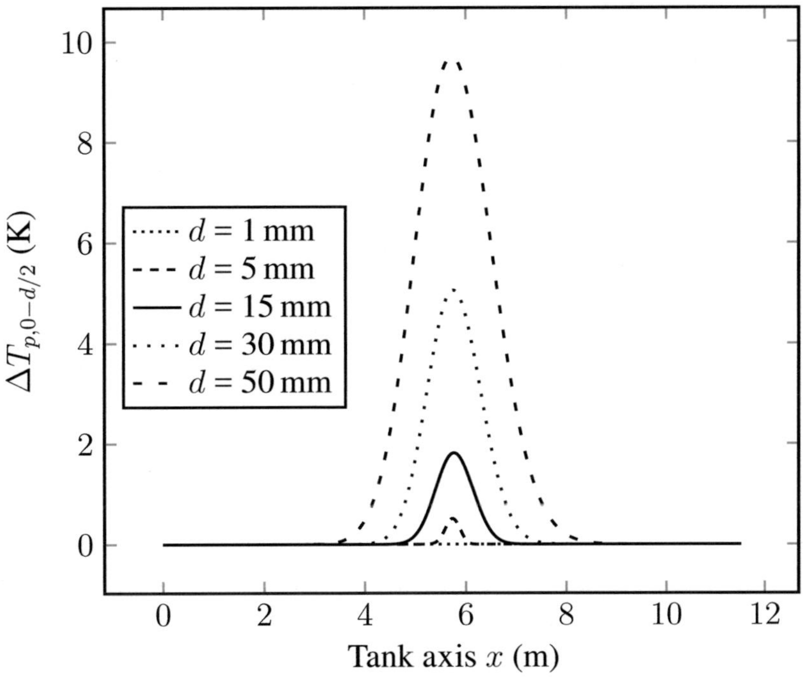

Figure 6.15: Temperature difference between core temperature ($y = 0$) and surface temperature ($y = d/2$) of solid particles after half of the discharge time (first cycle) along the axis of the tank for different filler diameters

A variation of the filler material influences the discharge efficiency, the standby behaviour, the storage-material cost and the pumping power as presented in Figure 6.16.

The useful discharged energy content increases slightly from quartzite (1) to iron (5) from 91.9% to 92.9% owing to the higher thermal conductivity in the filler particles. However, during standby this increased thermal conductivity is disadvantageous and the efficiency after a standby period of 8 h is reduced to 83.0% (5). During standby the thermocline thickness increases with increasing thermal diffusivity of the packed bed a_{mix} (Table 6.7) from 25.8% (1) to 45.1% (5).

Table 6.7: Storage tank height and thermal diffusivity a of the different considered filler materials in increasing order

Filler material	SiO_2	$MgAl_2O_4$	Al_2O_3	Aust. steel	Fe
no.	1	2	3	4	5
H (m)	11.5	11.3	11.0	10.0	9.8
$a_s = \frac{\lambda_s}{\rho_s c_{ps}}$ (10^{-6} m²/s)	0.9	1.3	1.5	4.8	17.7
a_{mix} (10^{-6} m²/s)	6.6	6.5	6.4	8.2	20.1

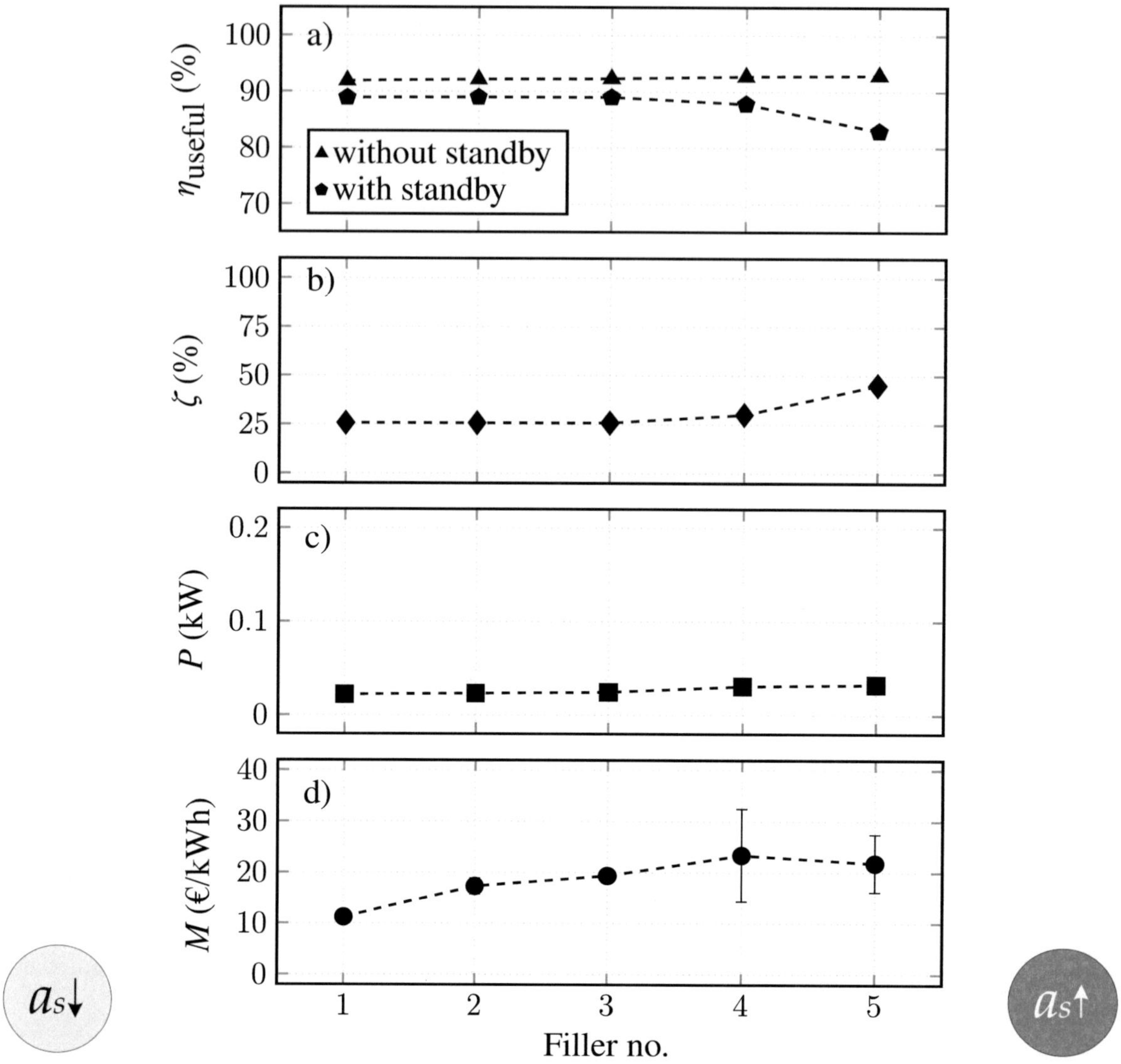

Figure 6.16: a) Discharge efficiency η_{useful} vs. filler material with increasing thermal diffusivity a_s; b) thermocline thickness after 8-h standby ζ vs. filler material; c) pumping power P vs. filler material; d) Storage-material cost M vs. filler material

In Figure 6.17, the dependency of the discharge efficiency and the thermocline degradation on the thermal conductivity λ_s and thermal capacity $\rho_s c_{ps}$ is shown separately. An increasing thermal conductivity of the filler material leads to a slight improvement of the discharge efficiency; however, if a standby is included, it leads to reduced efficiencies. Thus, a low thermal conductivity is favourable when sodium is used as the heat transfer fluid. This can also be seen in Figure 6.17b for austenitic steel (4) and iron (5), as their thermal capacities $\rho_s c_{ps}$ differ only slightly ($\approx 7\%$), but the four times higher thermal conductivity of iron leads to a drop of the discharge efficiency when a standby period is included.

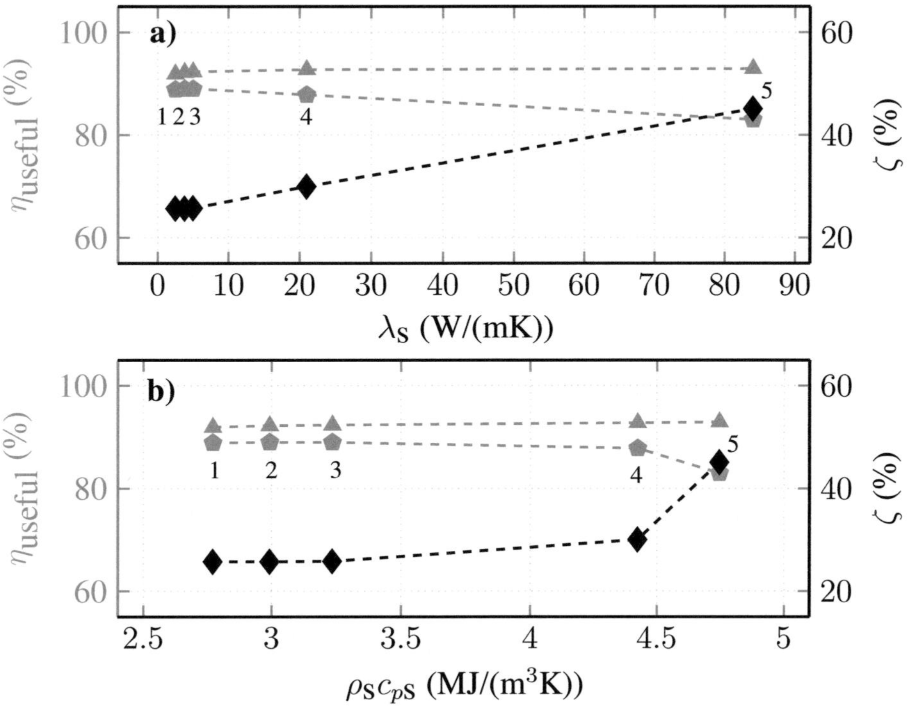

Figure 6.17: Discharge efficiency η_{useful} and thermocline thickness after 8-h standby ζ vs. a) filler material thermal diffusivity λ_s; b) thermal capacity $\rho_s c_{ps}$

The pumping power rises slightly from quartzite to iron owing to the decreasing tank diameter and thus, a slightly increased fluid velocity (Figure 6.16c. The lowest storage material costs are achieved when using quartzite (1) as filler material; with the other materials the costs are estimated to be approximately twice as high. However, the storage material cost of packed-bed storage systems with such a low porosity (ε = 0.22) are strongly dependent on the cost of specific cost of the solid filler material, which are given in Table A.6 in the Appendix. The error bars in Figure 6.16d show the spread of the costs. Therefore, it is difficult to draw a conclusion on the costs at this point.

6.3.5 Influence of velocity u_0

The velocity u_0 is calculated as shown in Equation 6.14. It can be varied either by changing the diameter of the tank, as done in Section 6.3.2, or by changing the mass flow. In this section, the tank diameter-to-height ratio is kept constant and the mass flow is varied (while keeping the overall storage capacity constant). Therefore, an increased velocity also implies a decreased storage/discharge time Δt according to Equation 6.19 (Table 6.8).

The velocity influences the discharge efficiency and the pumping power. The standby behaviour is not directly influenced; however, the different initial condition of the standby period (stable cycle) influences the size of the thermocline zone after 8-h standby. The results are shown in Figure 6.18.

Table 6.8: Influence of different velocities on mass flow and storage/discharge time

u_0 (mm/s)	0.5	1.0	1.9	3.5	5.0
$\dot{m}$ (kg/s)	10.4	20.9	39.7	73.1	104.5
Δt (h)	15.2	7.6	4.0	2.2	1.5
t_r (h)	1.4	0.7	0.4	0.2	0.1

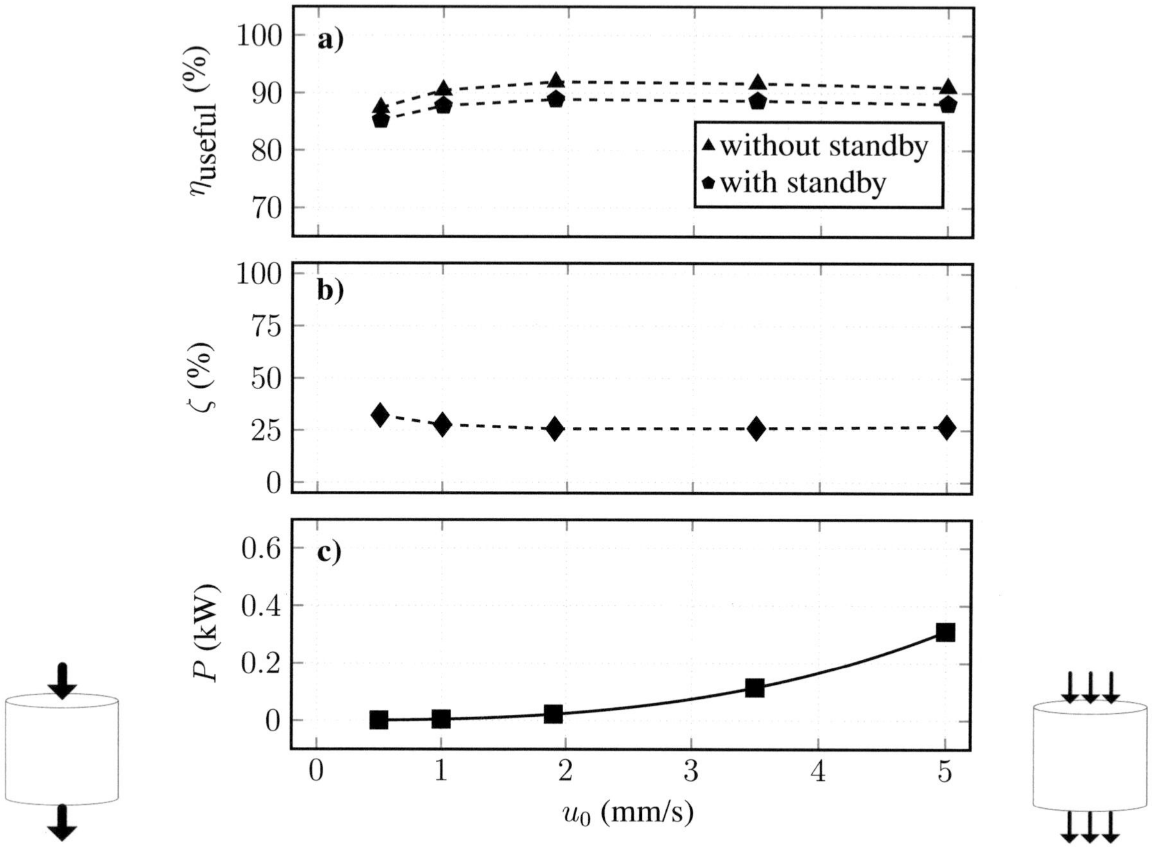

Figure 6.18: a) Discharge efficiency η_{useful} vs. velocity u_0; b) thermocline thickness after 8-h standby ζ vs. u_0; c) pumping power P vs. u_0

The influence of the fluid velocity on the discharge efficiency is negligible in the considered range and with the particle diameter of the reference case[5]. It is lowest (87.4%) for the minimum velocity

[5] In the Master thesis of M. Meßmer (2018), the influence on the velocity with different particle diameters is studied. The results show that for sufficiently small particles, an increasing velocity leads to a continuing increase of the discharge efficiency, as the thermal energy can be easily charged and discharged from small particles with a large heat transfer surface even for high velocities. For large particles, however, increasing the fluid velocity results in continuously decreasing discharge efficiencies. Now, the large particles and the comparatively small heat transfer surface limit the heat transfer and low velocities are necessary to charge and discharge the packed bed. With a medium diameter, the influence of the velocity is negligible and an optimal medium velocity leads to highest efficiencies, as it is the case in the present work.

(u_0 = 0.05 mm/s) and highest (91.9%) for the reference velocity (u_0 = 1.9 mm/s). This can be explained by two counteracting effects. On the one hand, an increase of the velocity means a decrease of the discharge time (Table 6.8) and hence, a shorter time for the thermocline zone to expand during discharge. On the other hand, higher velocities require a greater axial distance in the tank to fully heat the filler particles, resulting in an expanded thermocline zone. This is due to shorter residence times (Table 6.8). The residence time is defined in Equation 6.20 according to Ref. [148].

$$t_r = H \cdot \varepsilon / u_0 \tag{6.20}$$

Note that for all velocities, the heat transfer between fluid and filler is determined using the minimum Nusselt number, in the absence of a suitable correlation (see Section 5.1.3). In reality, especially for high velocities, the heat transfer would be improved and would lead to a better efficiency.

The thermocline thickness is smallest (25.7%) for the reference velocity (u_0 = 1.9 mm/s) and highest (32.1%) for the minimum considered velocity (u_0 = 0.05 mm/s). The degradation during standby is not dependent on the fluid velocity, but only on the condition after half of the (stable) discharge cycle, which is the initial condition for the 8-h standby period.

An increased velocity results in higher pumping power. However, even for the highest velocity of u_0 = 5 mm/s, the pumping power is below 0.008% of an electric output of 4 MW.

6.3.6 Influence of storage temperature

Three different storage temperature ranges are compared. Firstly, the reference case with 500–700 °C (T_{mean} = 600 °C), which is applicable to an advanced power cycle, and secondly, 290–565 °C (T_{mean} = 428 °C) suitable for a conventional power cycle in currently operational solar power plants. Those two cases include both a change of the temperature level (hence, different physical properties) and a change of the temperature difference (hence, different tank size). To separate those two effects, a third case is analysed with the same mean temperature as the reference case but the larger temperature difference of 275 K: 462.5–737.5 °C (T_{mean} = 600 °C). The changes in the mean storage temperature and temperature difference influence the discharge efficiency, the standby behaviour, pumping power and storage cost, as shown in Table 6.9.

Table 6.9: Resulting evaluation parameters for different operating temperatures

T_{mean} (°C)	428	600	
ΔT (K)	275	275	200
η_{useful} (%)	90.0	90.6	91.9
ζ (%)	30.8	30.2	25.7
η_{useful} incl. standby (%)	85.8	86.7	88.9
Pump. power P (W)	12.2	12.0	22.0
Cost M (€/kWh)	8.2	8.1	11.2

The discharge efficiency η_{useful} is slightly higher for a smaller temperature difference between maximum and minimum temperature in the tank. This higher efficiency results from the slightly higher storage tank, assuming the same D/H of the reference case (Table 6.10), which improves the discharge behaviour, as shown before. The different tank heights are caused by the varied temperature difference while the storage capacity is constant. Additionally, the thermal conductivity λ is lower at higher temperatures, leading to smaller thermocline zones improving the discharge efficiency. A lower thermal conduction also leads to a reduced heat transfer between fluid and filler at the same time (Table 6.10) but it is not the dominating factor.

Table 6.10: Characteristic parameters influencing the storage behaviour at different temperatures

T_{mean} (°C)	428	600	
ΔT (K)	275	275	200
H (m)	10.4	10.4	11.5
λ (W/(mK))	70.8	62.4	
h_v (10^6 W/(m^{3}K))	2.95	2.60	
a_{mix} (10^{-6} m^2/s)	7.3	6.6	

If a standby period is included during discharge, the efficiency is reduced more strongly for the cases with a higher temperature difference owing to the smaller tank heights compared with the reference case. Additionally, the higher thermal conductivity of the fluid at lower temperatures and thus, a higher overall thermal diffusivity a_{mix} in the packed bed (Table 6.10), results in an increased thermocline degradation during standby for the lower temperature case.

The required pumping power is lower for higher temperature differences owing to both a reduced tank height and a reduced mass flow. These outweigh the counteracting influence of the smaller tank diameter on the pressure loss. The storage cost is reduced for higher temperature differences in the tank (from 11.2 €/kWh to 8.2 €/kWh), as a smaller amount of storage material is necessary to provide the storage capacity.

6.3.7 Conclusion

In Table 6.11 the investigated parameters and their effect on the selected performance criteria are summarized. A low porosity, a small tank diameter-to-height ratio and small filler particles are advantageous and lead to highest discharge efficiencies and minimum thermocline degradation during standby. The pumping power, however, increases with all the mentioned parameters decreasing, but remains at very low absolute values. The storage medium costs are not influenced by tank diameter-to-height ratio and filler particle size, while reduced porosities lead to lowest costs (9.7 €/kWh). As the target of 15 €/kWh is not likely to be met when additional costs such as piping, insulation etc. are considered, further cost reductions are necessary that can be achieved only with extremely low cost filler material.

The optimal filler material would be one with low thermal conductivity and a high thermal capacity ρc_p. The discharge efficiencies without standby are not significantly influenced by the choice of filler material; however, if a standby period is included, the filler material with high thermal conductivity (iron) shows a reduced discharge efficiency. Using quartzite as filler material results in lowest storage-material costs.

The velocity has no significant influence on the discharge efficiency with the considered particle diameter. For smaller diameters, however, an increasing velocity leads to an increasing dischage efficiency, and for large diameters to a decreasing discharge efficiency. The pumping power is highest for highest fluid velocities, but still in a range below 0.01% of an electric output of 4 MW.

A reduced temperature difference in the tank leads to greater tank heights when the storage capacity is kept constant, resulting in improved efficiencies. However, this implies a larger quantity of storage material required and hence, higher storage medium costs. Furthermore, higher mean operating temperatures are beneficial owing to the lower thermal conductivity of sodium at higher temperatures.

In the following, results with optimised storage parameters are presented. For this case, a porosity of $\varepsilon = 0.1$, a ratio of $D/H = 0.2$ and a filler diameter of $d = 5$ mm is selected. With these parameters, the tank height (20.7 m) is still feasible and the pumping power is below 1% of the electric output. The filler material and temperature range are used as as given in the reference case (Section 6.2).

The storage efficiency η_{useful} is 96.4% for the selected optimal parameters, which is $\approx 4\%$ higher than the results from the reference case (Table 6.12). During standby the thermocline zone in creases to 11.3% (2.3 m), which is lower than the lowest value obtained in the parametric study. The standby period reduces the discharge efficiency to 95.4%. The pumping power of 7 kW is $< 0.2\%$ of a possible electrical power of 4 MW. The storage cost is lowest due to the highest amount of filler in the packed bed considered in the parametric study.

Table 6.11: Overview of results of parametric study; ↑↑: large increase, ↑: increase, (↑): minor increase; ↓↓: large decrease, ↓: decrease, (↓): minor decrease

Parameter	η_{useful}	ζ	P	M	Result
$\varepsilon\downarrow$	↑↑	↓↓	(↑)	↓↓	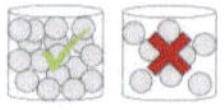
$D/H\downarrow$	↑↑	↓↓	(↑)	–	
$d\downarrow$	↑↑	↓	↑	–	
Filler material $a_s\downarrow$	↑	↓	(↓)	↓↑	
$u_0\uparrow$	(↑↓)	(↓↑)	(↑)	–	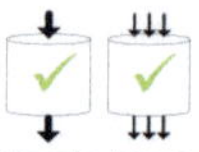
$\Delta T\downarrow$	(↑)	(↓)	(↑)	↑	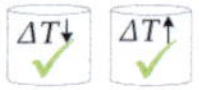

Table 6.12: Evaluation parameters with optimized storage parameters $\varepsilon = 0.1$, $D/H = 0.2$ and $d = 5$ mm compared to results with reference parameters (Section 6.2.3)

	Optimised	Reference
η_{useful} (%)	96.4	91.9
ζ (%)	11.3	25.5
η_{useful} incl. standby (%)	95.4	88.9
Pump. power P (W)	7.0	22.0
Cost M (€/kWh)	9.7	11.2

6.4 Parametric studies for conventional fluids

A comparison with parametric studies of packed-bed storage systems with conventional heat transfer fluids (salt mixtures, oils) shows similar trends for the dependency of the discharge efficiency on the D/H ratio and the filler diameter, as shown in Refs. [86, 87, 89, 90, 91]. The influence of different filler material on the discharge efficiency of molten salt in a packed-bed storage is analysed by Xu et al. [90], and they conclude that for a sufficiently small particle size the performance is not influenced by the choice of filler material. Only for large diameters (> 100 mm) is an influence apparent. The studies in this work, by contrast, show an influence of the storage material even with a smaller particle diameter of $d = 15$ mm, particularly in the case of the highly conductive iron, which leads to poor standby behaviour. The influence of the velocity on the discharge efficiency of a molten salt packed bed is as observed for sodium in this work [89]. Furthermore, no significant influence of the variation of the inlet temperature (while keeping the maximum temperature constant) is reported by Xu et al. [89] for molten salts; however, the authors keep the tank height constant and not the thermal capacity. Thus, no influence of the tank height is present.

Interestingly, a completely opposite trend for storage efficiency for different bed porosities is seen when molten salts are used. The efficiency is higher for high porosities, owing to the better storage properties (high ρc_p, low λ) of the salts compared with the filler material and the poor heat transfer between fluid and filler [176]. The results of Xu et al. [89] agree with these trends on the efficiency. Van Lew et al. [87] and Reddy et al. [88], however, obtain an optimal efficiency for porosities between 0.15 and 0.35, because they use a Nusselt correlation which is dependent on the porosity. Again, this Nusselt correlation is not applicable for low Prandtl numbers and thus is not used for sodium in this study.

Yang and Garimella [85] propose a correlation for the influence of the Reynolds number ($\mathrm{Re_p}$) and the ratio of the tank height to the filler diameter (H/d) based on their simulation results for molten salt packed-systems with a given porosity. Figure 6.19 shows that this correlation is not able to represent the simulated data from Section 6.3 when using sodium in the packed-bed storage, as the influence of the Prandtl number is not considered. A large database beyond the scope of this work is necessary to develop a correlation for the discharge efficiency as a function of $\mathrm{Re_p}$, Pr, H and ε, and thus to predict the numerical results.

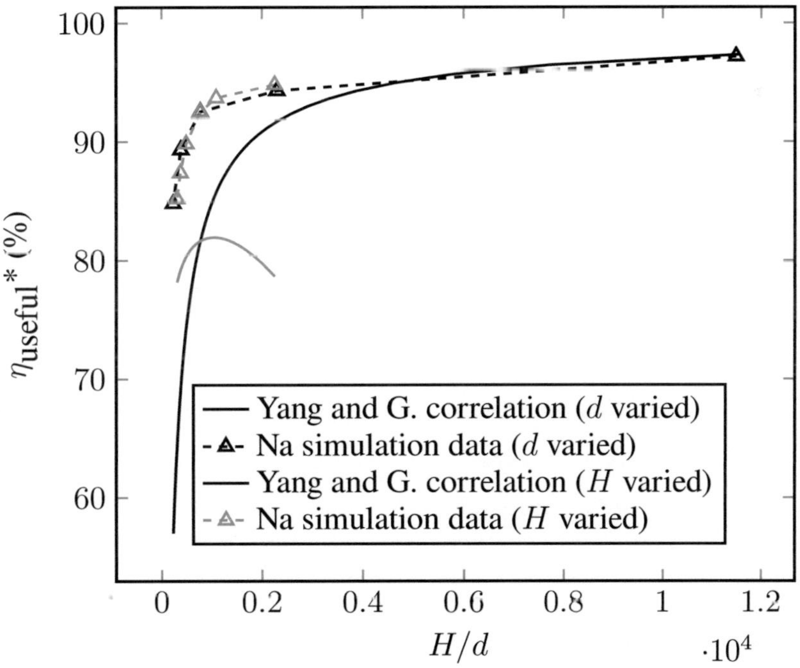

Figure 6.19: Comparison of discharge efficiency in the first discharge step (*with a threshold of 10 K as defined in Ref. [85]) from the correlation of Yang and Garimella [85] and simulated data from this work

6.5 Comparison with other fluids

The performance of sodium in a packed-bed thermocline storage is compared with the state-of-the-art fluid solar salt at temperatures of 290–565 °C and high-temperature salts at 500–700 °C. The storage parameters of the reference case (Section 6.2) are taken for this comparison with $\varepsilon = 0.22$, $D/H = 0.5$ and $d = 15\,\text{mm}$. The influence of the physical properties is compared and this time no optimization is performed. Additionally, the performance of lead-bismuth eutectic (LBE) is investigated. LBE is tested at KIT as a possible heat transfer fluid in a thermal receiver, as it also has a high thermal conductivity (more than 20 times higher than solar salt), a large temperature range and easier handling than sodium [179]. Studies showed a potential LCOE reduction of 6% for LBE compared with solar salt in a CSP receiver loop at conventional temperatures [7]. Lead is only included in the comparison at 500–700 °C, owing to its melting temperature of 327 °C. Its specific material cost is approximately one tenth that of LBE; however, its use requires extensive heat tracing because of its relatively high melting temperature.

6.5.1 Sodium vs. solar salt, LBE

The use of sodium and solar salt in a packed-bed storage is compared at 290–565 °C. The physical properties of sodium, LBE and solar salt can be found in the Appendix (Section A.3). The raw material cost of solar salt and LBE is 1.0 €/kg and 12 €/kg, respectively [44]. The storage parameters are presented in Table 6.13.

Table 6.13: Storage parameters for each fluid (with physical properties at T_{max})

	Na	LBE	Solar salt
Height H (m)	10.4	10.3	9.9
Fluid mass m_{f} (tons)	39.8	465.3	72.5
Solid mass m_{s} (tons)	451.1	436.1	392.3
Mass flow $\dot{m}$ (kg/s)	29.0	257.5	23.6
Velocity u_0 (mm/s)	1.7	1.2	0.7

The storage tank is largest for sodium, as it has the lowest thermal capacity ρc_p. The low density leads to the least required fluid mass for sodium compared with the other fluids. The fluid mass of LBE, by contrast, is ≈ 10 times higher than sodium owing to its ≈ 10 times higher density. The high mass flow for LBE is due to its low specific heat capacity (Equation 6.19).

In Figure 6.20 (left) the temperature distributions of sodium, LBE and solar salt along the tank axis in a stable discharge cycle are presented. The fluid temperature of LBE and sodium have the sharpest curvature, leading to a longer time in which the fluid can be extracted at maximum temperature at the outlet of the tank, as shown in Figure 6.20 (right). Both liquid metals show a smaller thermocline region owing to the higher heat transfer between fluid and filler compared with the solar salt packed bed.

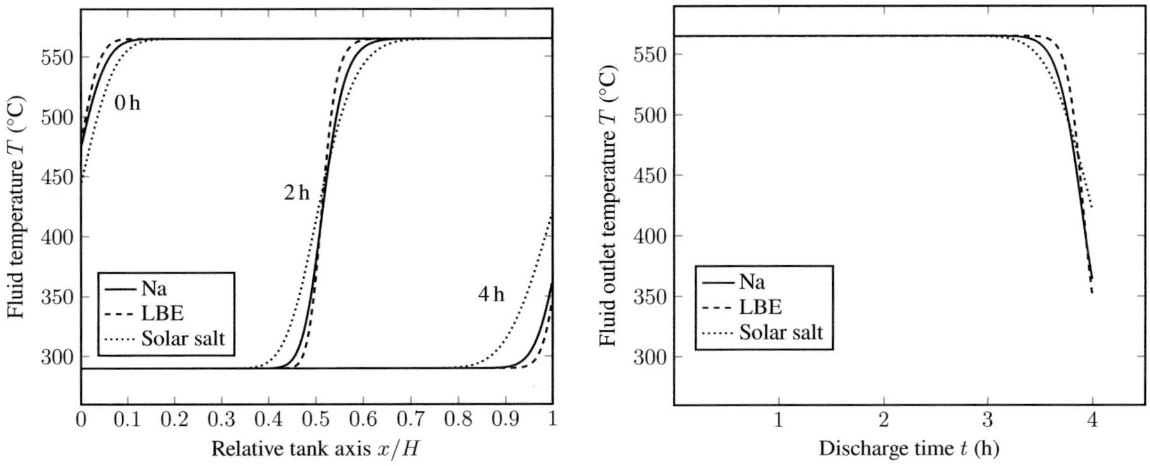

Figure 6.20: Temperature distributions of Na, LBE and solar salt; left: fluid temperature along relative tank axis at 0 h, 2 h, and 4 h discharge time intervals in a stable cycle; right: fluid outlet temperature vs. discharge time

The resulting efficiency parameters, pumping power and storage cost are presented in Table 6.14. The discharge efficiency η_{useful} is highest for liquid metal packed-bed storage systems by reason of their excellent heat transfer characteristics. LBE produces the highest efficiencies, as the thermal conductivity is lower than for sodium, leading to a sharper temperature stratification in the tank while maintaining an excellent heat transfer between fluid and filler at the same time. The volumetric heat transfer coefficient h_{v} (at T_{max}) between sodium and the filler particles is 2.7 MW/(m^{3}K), for LBE it is 0.6 MW/(m^{3}K), and for the solar salt packed bed it is 0.02 MW/(m^{3}K) . During standby, the high thermal conductivities of the liquid metals (λ_{Na} = 64.1 W/(mK), λ_{LBE} = 15.1 W/(mK)) lead to a large extension of the thermocline zone; from 16.6% to 30.8% for sodium and from 11.4% to 17.7% for LBE. The thermocline zone of the solar salt packed bed, in contrast, expands only from 25.2% to 26.8%. As a result, the discharge efficiency including an 8-h standby period is reduced more significantly for the liquid metal packed beds than for the solar salt packed bed; 85.8% for sodium, 90.9% for LBE and 86.9% for solar salt.

The highest pumping power is required in the LBE storage by reason of its high mass flow and high density. The storage cost of the solar salt packed bed is lowest owing to the low-cost storage fluid; the sodium storage is 22% more expensive. The LBE packed-bed storage is more than 17 times more expensive than the sodium packed-bed storage because of the high raw-material cost of LBE.

Table 6.14: Results for sodium, LBE and solar salt at 290–565 °C

	Na	LBE	Solar salt
η_{useful} (%)	90.0	92.8	87.4
ζ (%)	30.8	17.7	26.8
η_{useful} incl. standby (%)	85.8	90.9	86.9
Pump. power P (W)	12.2	50.5	5.2
Cost M (€/kWh)	8.2	145.0	6.7

6.5.2 Sodium vs. high-temperature salts, LBE, lead

The performance of sodium at 500–700 °C is compared with that of three high-temperature salts proposed in the Concentrating Solar Power Gen3 Demonstration Roadmap [4]: $ZnCl_2$-NaCl-KCl (composition by weight 0.686-0.075-0.239), $MgCl_2$-KCl (composition by weight 0.375-0.625) and Na_2CO_3-K_2CO_3-Li_2CO_3 (composition by weight 0.334-0.345-0.321). They are named HTS 1, HTS 2 and HTS 3 in the following. The physical properties and cost data of HTS 1–3 can be found in the Appendix (Section A.3 and A.4). The results are published in Ref. [176] and summarized in this work. Additionally, a comparison with LBE and lead is performed here. The resulting storage parameters for each fluid are presented in Table 6.15. The differences between sodium and LBE and the molten salts can be explained as in Section 6.5.1.

Table 6.15: Storage parameters for each fluid (with physical properties at T_{max})

	Na	LBE	Pb	HTS 1	HTS 2	HTS 3
Height H (m)	11.5	11.4	11.4	11.3	11.2	10.9
Fluid mass m_{f} (tons)	53	631	651	123	102	104
Solid mass m_{s} (tons)	622	602	598	581	574	526
Mass flow $\dot{m}$ (kg/s)	39.8	359	352	55.6	43.5	31.0
Velocity u_0 (mm/s)	1.9	1.4	1.4	1.1	1.1	0.7

The fluid temperature distributions along the tank axis and the outlet temperature over the discharge time of sodium, LBE and HTS 1 are compared in Figure 6.21. One high-temperature salt is presented exemplarily for all three investigated salts, as the temperature distributions do not differ significantly, as shown in Ref. [176]. The fluid temperature of the liquid metals again show a smaller thermocline zone in Figure 6.21 (left), and the fluid can thus be extracted at the maximum temperature at the top of the tank for a longer time during discharge.

The performance parameters are shown in Table 6.16. The discharge efficiency η_{useful} is highest when liquid metals are used, owing to the high heat transfer rate between fluid and solid material.

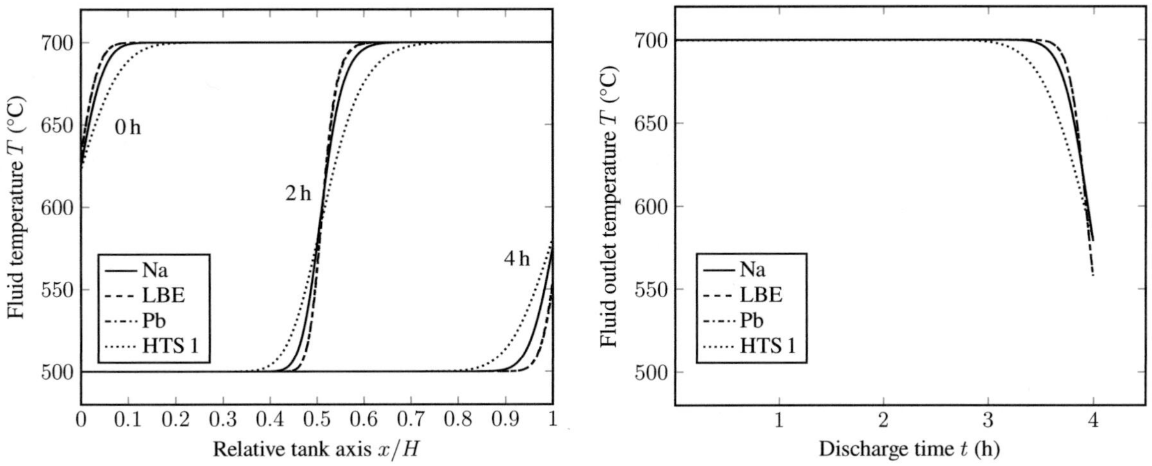

Figure 6.21: Temperature distributions of Na, LBE, Pb and HTS 1; left: fluid temperature along relative tank axis at 0 h, 2 h, and 4 h discharge time intervals in a stable cycle; right: fluid outlet temperature vs. discharge time

Again, LBE (and lead) shows the highest efficiencies. During 8 h standby, the thermocline zone of the sodium packed bed increases from 15.2% to 25.7%, leading to a reduction of 2% in the discharge efficiency including standby. The thermocline zone of the LBE storage expands from 10.5% to 16.0%, resulting in a reduced efficiency of 92.2% after 8-h standby. For lead similar degradations are obtained. When using the high-temperature salts, ζ increases only $\approx 1\%$ during standby and therefore, the discharge efficiency is only slightly reduced by 0.4%.

Table 6.16: Results for each fluid at 500–700 °C

	Na	LBE	Pb	HTS 1	HTS 2	HTS 3
η_{useful} (%)	91.9	93.7	93.7	86.0	87.5	87.0
ζ (%)	25.7	16.0	16.6	27.4	25.4	26.2
η_{useful} incl. standby (%)	88.9	92.2	92.0	85.6	87.1	86.6
Pump. power P (W)	22.0	96.0	90.6	63.5	63.9	31.0
Cost M (€/kWh)	11.2	197.0	33.5	11.2	8.2	13.3

The required pumping power is lowest for sodium compared with the salts owing to its low mass flow, low viscosity and low density. Pumping LBE and lead demands the highest pumping power due to their high mass flow and high density. The overall storage-material cost is lowest for the HTS 2 storage, as this salt mixture has the lowest cost (0.4 €/kg). The LBE storage-material costs are more than 17 times higher than those for the sodium packed-bed storage, which makes this storage system uninteresting from an economic point of view. The storage material for a packed-bed storage with lead is three times as expensive as with sodium owing to its high required mass resulting from its low specific heat capacity.

Be reminded that no optimization has been performed for all fluids. For salts, firstly, high porosities are favourable regarding the efficiency and secondly, even smaller particles sizes could lead to similar discharge efficiencies as obtained with liquid metals, as shown for high-temperature salts by Niedermeier et al. [176] and for solar salt in the Master thesis of M. Meßmer (2018). However, a low porosity is desired for packed-bed storage systems for all fluids due to low storage-material costs and the filler diameter should not lead to immense required pumping powers. Thus, the comparison with the selected storage parameters is suitable.

6.6 Conclusion

A parametric study was performed optimizing the discharge efficiency η_{useful}, the degradation of the thermocline during standby, the pumping power and the storage-material cost. Changing the porosity, the D/H ratio and the filler particle size showed the strongest influence on the performance of the packed-bed storage. Highest discharge efficiencies and best standby behaviour were reached for low porosity and D/H ratio and small filler particles. The cost of the storage material needed was also minimum for low porosities, but needs to be further reduced by selecting even less expensive filler material to meet the desired cost target. The pumping power remained low in all cases. Therefore, a packed-bed thermal storage with sodium should be packed with solid particles as densely as possible. To achieve porosities below 0.26 a multidisperse packing is necessary, while keeping the mean diameter of the filler as low as possible.

The comparison with solar salt at conventional temperatures (290–565 °C) showed the advantages of sodium during cycling. Thanks to its high thermal conductivity, the heat can be transferred efficiently to and from the filler particles. The discharge efficiency η_{useful} was 2.6% higher for sodium (90.0%) than for solar salt (87.4%). However, if a standby period was included in the discharge step, the efficiency was reduced to 85.5% for sodium owing to the high conduction in the packed bed. For solar salt, the discharge efficiency was reduced only to 86.9%. The storage-material costs were 8.2 €/kWh for the sodium packed bed compared with 6.7 €/kWh for solar salt, both using quartzite as filler material. A further comparison with salt mixtures suitable for higher temperatures (500–700 °C) presented the same trend. During discharge, sodium showed a discharge efficiency η_{useful} of 91.9% compared with 86.0 to 87.5% for the salts. The required pumping power and storage-material costs were in the same range for sodium and the considered salts.

In conclusion, sodium can be competitive with salts at elevated temperatures. Even higher discharge efficiencies can be reached. To counteract its faster thermocline degradation during standby, technical solutions can be applied or short standby times can be used for a packed-bed storage with sodium. The liquid metal LBE showed best efficiencies as it also has a high thermal conductivity, leading to an excellent heat transfer between filler and fluid but also to a better stratification thanks to its lower thermal conductivity than that of sodium. However, the storage-material costs were more than 10 times higher than for a packed bed with sodium, making LBE unattractive as a storage medium. With lead, high efficiencies similar to those with LBE were obtained, but the storage-material costs were still three times higher than with sodium.

Based on the parametric study presented in this section, an arrangement of multiple tanks will be investigated in the next section which may allow for both high discharge efficiencies owing to low D/H ratios with realistic tank heights and improved standby behaviour thanks to the thermal separation of the different temperature zones in individual tanks.

7 Performance of sodium in a multi-tank packed-bed storage

The principle of the multi-tank arrangement is splitting a single tank into multiple tanks in serial connection while maintaining the same storage capacity (Figure 7.1). A lower D/H ratio can be realized in a multi-tank arrangement than with a single tank with the same storage capacity. Additionally, the thermal conduction in axial direction in a packed bed with sodium as heat transfer fluid – which leads, particularly during standby, to a fast degradation of the thermocline region – is interrupted by splitting the single tank into multiple tanks. In principle, it amounts to inserting insulation in several axial layers in a single tank. The disadvantage is the additional costs due to more required insulation, instrumentation and piping and an increased required pumping power.

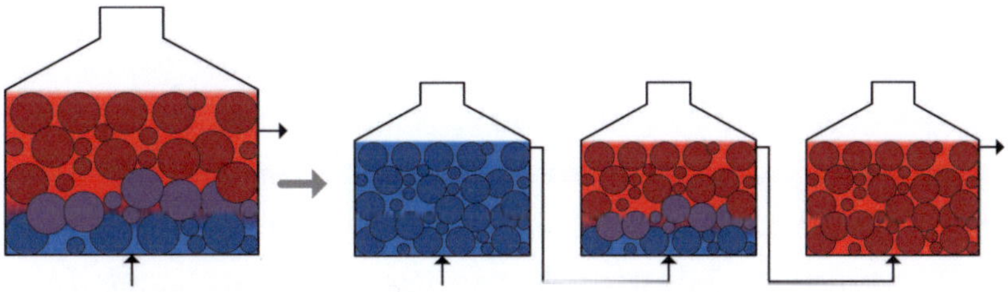

Figure 7.1: Scheme of a discharge process in a single-tank compared with a multi-tank arrangement

Publications about using multiple packed-bed or thermocline storage tanks are scarce. Fritsch et al. [44] suggest a segmented packed-bed storage system for liquid metal and perform a cost analysis. For sodium, the single-tank system with quartzite is less than half as expensive as the segmented storage due to additional costs for insulation etc. Angelini et al. [94] perform parametric studies for a molten salt packed-bed storage. They suggest two serial thermocline tanks as a consequence to enable small D/H ratios and recommend further studies. Osorio et al. [180] conduct an exergy analysis of discharging a multi-tank thermal energy storage system applicable to the operating temperatures of water, oils and molten salts (without filler material). They found that the efficiency can be improved by splitting the storage system into multiple tank with negligible pumping and heat loss penalties. For the application in solar heating systems with water as storage medium, multi-tank systems are studied both experimentally and numerically. Charging and dicharging efficiencies are evaluated using parallel and serial connections [181, 182]. For air as heat transfer fluid, segmented/modular storage systems are investigated to enhance the stratification and to get a flexible and efficient air flow design [183, 184, 185].

7.1 Modifications of the model

In the multi-tank arrangement, the fluid flows through the tanks in serial connection, as shown in Figure 7.1 for discharge. For charging, the fluid enters the last tank from the top and leaves at the bottom to enter at the top of the previous tank. The temperature distribution in each tank is calculated as for a single tank, as explained in Section 5. Only the boundary condition for the fluid at the inlet of the tank is changed. For the first tank ($k = 1$), it is a constant value T_{in}, but for the others it is the outlet temperature of the previous tank T_{out}.

$$x = 0: \quad \begin{cases} T(k) = T_{\text{in}} & (k = 1) \\ T(k) = T_{\text{out}}(k-1) & (k > 1) \end{cases} \tag{7.1}$$

The remaining boundary conditions are unchanged (Section A.5.4). It is assumed that the tanks do not influence each other from tank wall to tank wall by heat exchange.

7.2 Methodology

The performance of sodium in a multi-tank arrangement is compared with that in a single tank by keeping the overall storage capacity constant. The storage tank height H is determined in the following equation by dividing the overall storage capacity Q by the number of storage tanks n_{tanks}. The fluid and solid mass and the fluid mass flow are dependent only on the overall capacity and discharge time and are the same in a single- and a multi-tank arrangement.

$$H = \left(\frac{4Q/n_{\text{tanks}}}{\pi (D/H)^2 (\rho_{\text{f}} c_{p\text{f}} \varepsilon + \rho_{\text{s}} c_{p\text{s}} (1-\varepsilon)) (T_{\max} - T_{\min})} \right)^{1/3} \tag{7.2}$$

The performance parameters are determined in a way analogous to that used for the single-tank storage system (Section 6.1) with the following adaptations to the multi-tank system: The discharge efficiency η_{useful} is defined using the outlet temperature of the last tank. To assess the standby behaviour, the height of the thermocline region is related not only to the height of one tank, but also to the total tank height of all tanks. The pumping power is the sum of the required pumping power for each tank. The storage cost still represents only the bulk material cost, and thus remains unchanged. Additional insulation, piping and instrumentation costs are expected to be higher than for the single-tank arrangement and hence have to be kept in mind when comparing single- with multi-tank arrangements.

The same 40 MWh reference case as for the single-tank study is used (Section 6.2) to allow a direct comparison. A parametric study is performed, where two parameters are varied: the number of tanks n_{tanks} and the D/H ratio (Table 7.1). The influence of the porosity ε and filler diameter d are not investigated, as their influence on the performance in a packed-bed storage shown for a single tank can be transferred to a multi-tank system.

Table 7.1: Varied parameters for the multi-tank system, their range and the affected performance parameters (reference case bold)

Parameter	Range	η	ζ	P	M
D/H	[0.1,**0.5**,2.0]	✓	✓	✓	–
n_{tanks}	[**1**,3,5,7,9,15,20]	✓	✓	✓	–

7.3 Results

The results for different numbers of tanks and with different D/H ratios are presented in the following. Further, the performance of sodium in a multi-chamber arrangement and the applicability of a sodium multi-tank configuration to a state-of-the-art storage size is tested.

7.3.1 Influence of number of tanks

An increased number of tanks at a constant D/H ratio lead to reduced tank heights, higher velocities and shorter residence times in the tanks (Table 7.2).

Table 7.2: Tank height H, velocity u_0 and resulting residence time $t_r = H \cdot \varepsilon / u_0$ for different number of tanks

n_{tanks}	1	3	5	7	9	15	20
H (m)	11.5	8.0	6.8	6.0	5.6	4.7	4.3
u_0 (mm/s)	1.9	4.0	5.6	7.0	8.2	11.6	14.0
t_r (s)	1.3	0.4	0.3	0.2	0.2	0.1	0.1

The number of tanks has an influence on the discharge efficiency, the thermocline degradation during standby and the required pumping power, as shown in Figure 7.2.

The discharge efficiency η_{useful} increases with an increasing number of tanks. The efficiency improves from 91.9% (single tank) to 93.8% (three tanks); for five tanks it increases to 94.1%. A further increase of tank numbers does not significantly improve the discharge efficiency. For 20 tanks, 94.6% efficiency is reached. The main reason for an increase of the efficiency is that the thermal conduction in axial direction is prevented between each storage tank. At the same time also the total tank height, which is flown through by the fluid, increases, which impoves

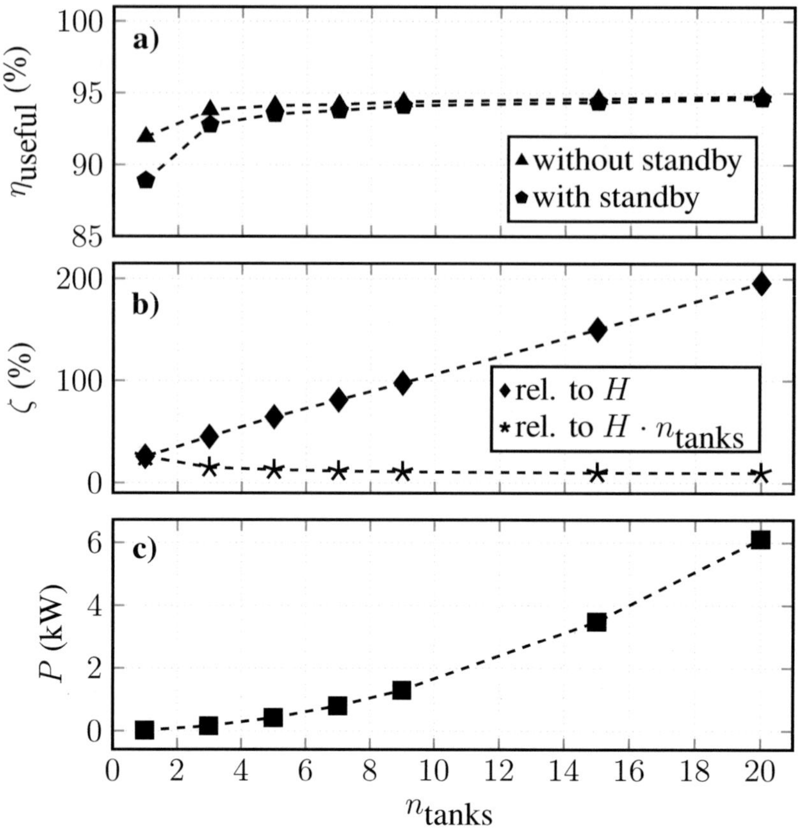

Figure 7.2: a) Discharge efficiency η_{useful} vs. number of tanks n_{tanks}; b) thermocline thickness after 8 h standby ζ vs. n_{tanks}; c) pumping power P vs. n_{tanks}

the discharge behaviour as shown earlier for the single tank. However, an increasing number of tanks also means an increased velocity and smaller heights of each tank, resulting in a shorter residence time as shown in Table 7.2. As a result, a larger share of the tank height is necessary to fully heat/discharge the particles, as already shown for the single-tank in Section 6.3.5. These counteracting effects lead to a limitation of the efficiency increase.

The influence of an included standby period on the discharge efficiency is negligible when the number of tanks is increased to more than three. This is due to the fact that, during standby, only the tanks occupied by the thermocline zone are affected. With a single-tank configuration, the temperature in the complete tank is influenced. For three to nine tanks, the thermocline is limited to the central tank after half of the stable discharge step. This is exemplarily shown in Figure 7.3 for a single- and a five-tank arrangement. The temperature in the packed bed affects only the central one of the five tanks (from 0.4 to 0.6 of the total tank heights), and the thermocline degradation after 8 h of standby is reduced severely compared with that in the single tank.

The greater the number of tanks, the more likely it is that the thermocline region occupies more than one tank. For 15 tanks, the thermocline region extends to two tanks after half of the stable discharge step and thus, two tanks are affected by the thermal conduction during standby. The

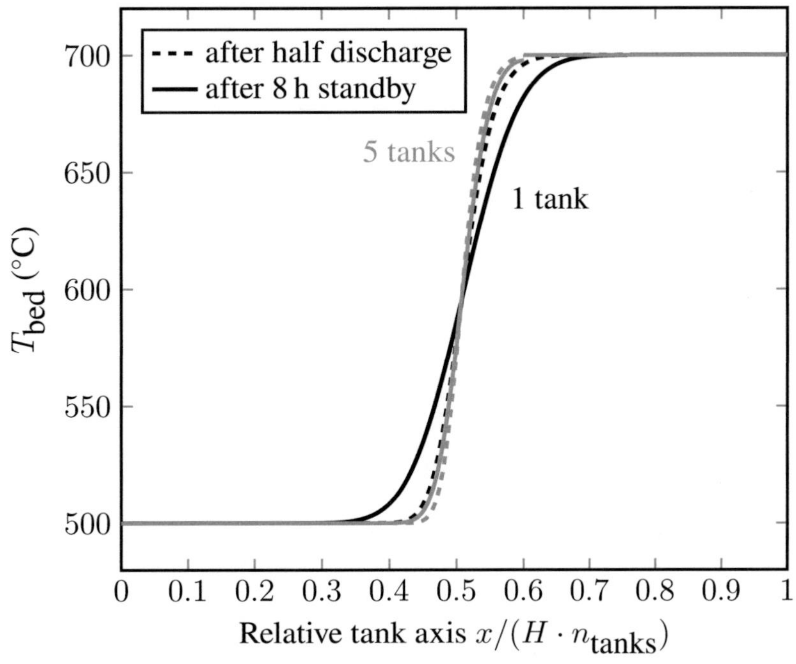

Figure 7.3: Temperature distribution in the packed bed along the axis of the tank(s) after half of a discharge cycle and after an additional 8-h standby period

thermocline thickness related to the tank height reaches 150%. When 20 tanks are used, the thermocline expands to three tanks during discharge and the thermocline related to the tank height reaches ≈ 200%. The thermocline thickness related to the sum of all tank heights in Figure 7.2 b emphasizes the benefit of multi-tank configurations. For example, with seven tanks the thermocline zone reaches H_{thcl} = 4.9 m, which is 81.3% of one of the seven tanks and 11.6% of the total tank height. However, similar to the discharge efficiency, only small improvements are obtained by increasing the number of tanks to more than five for the reference case selected in this study.

For longer standby periods, however, a large number of tanks is always advantageous, as the thermocline eventually occupies the complete tank at a certain standby time, as shown in Figure 7.4. In case of a single tank the thermocline expands throughout the entire storage volume. In a five-tank arrangement, only one fifth of the overall storage volume (the central tank) is occupied by the thermocline zone after more than 2 days, that is, if a standby period occurs after discharging half of the storage. If starting at other points in time during discharge, the thermocline is present in more than one tank during standby.

When 20 tanks are used, the thermocline expands in two tanks and thus results in 10% of the overall tank volume when leaving the tanks in standby mode. A further tank holds a small part of the thermocline region at the bottom of the tank. In this case, however, the temperature in the whole tank adjusts to a temperature above $T_{\text{max}} - 5$ K during standby, leading to a decrease in ζ in this tank. Thus, the overall $\zeta*$ has a maximum higher than 10% at ≈ 2 days. The curves of the thermocline thickness (Figure 7.4) show a peak when the thermocline reaches the top of the tank ($T_{\text{max}} - 5$ K). When it also extends to the bottom of the tank, the maximum $\zeta*$ is reached.

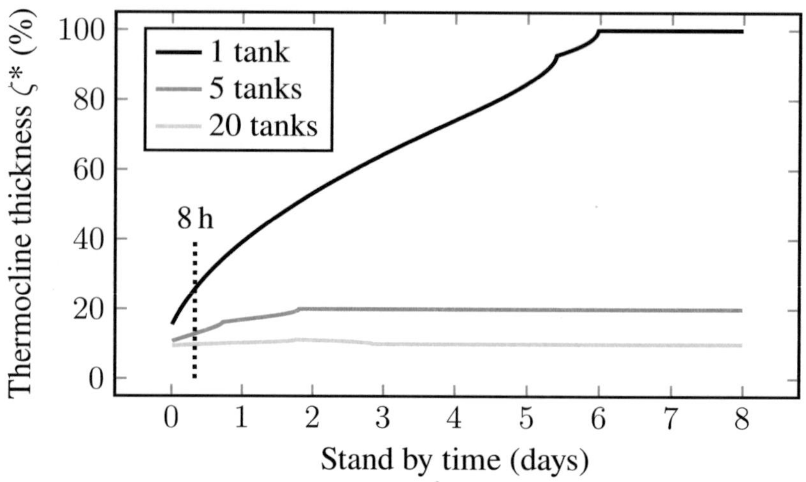

Figure 7.4: Degradation of the thermocline during standby starting from half of a discharge step in a stable cycle, *related to total tank height ($H \cdot n_{\text{tanks}}$)

The pumping power increases with increasing number of tanks due to an increased flow length and to higher velocities in the tanks. The latter result from smaller tank diameters at constant mass flow. The maximum required pumping power of 6.1 kW is still < 0.2% of an electrical output of 4 MW. The Ergun equation for calculating the pressure loss in the multi-tank arrangements is valid according to Stieß [172] with Reynolds numbers $\text{Re}_\text{p} < 900$.

7.3.2 Influence of D/H ratio

The D/H ratio is varied from 0.1 to 2.0 for selected numbers of tanks, again maintaining the overall storage capacity. An overview of the varied parameters, the resulting tank geometries and velocities is presented in Table 7.3.

Table 7.3: Tank geometries, mass flow and velocity for different numbers of tanks and D/H ratios

D/H		0.1			0.5			2.0	
n_{tanks}	1	5	20	1	5	20	1	5	20
H (m)	33.8	19.7	12.4	11.5	6.8	4.3	4.6	2.7	1.7
D (m)	3.4	2.0	1.2	5.8	3.4	2.2	9.2	5.4	3.4
u_0 (mm/s)	5.6	16.3	41.0	1.9	5.6	14.0	0.8	2.2	5.6

The D/H ratio influences the discharge efficiency, the standby behaviour and the required pumping power. The results are shown in Figure 7.5.

The discharge efficiency η_{useful} is strongly influenced by the D/H ratio in a single-tank arrangement and best for small D/H ratios: 94.4% ($D/H = 0.1$) compared with 83.9% ($D/H = 2.0$). This

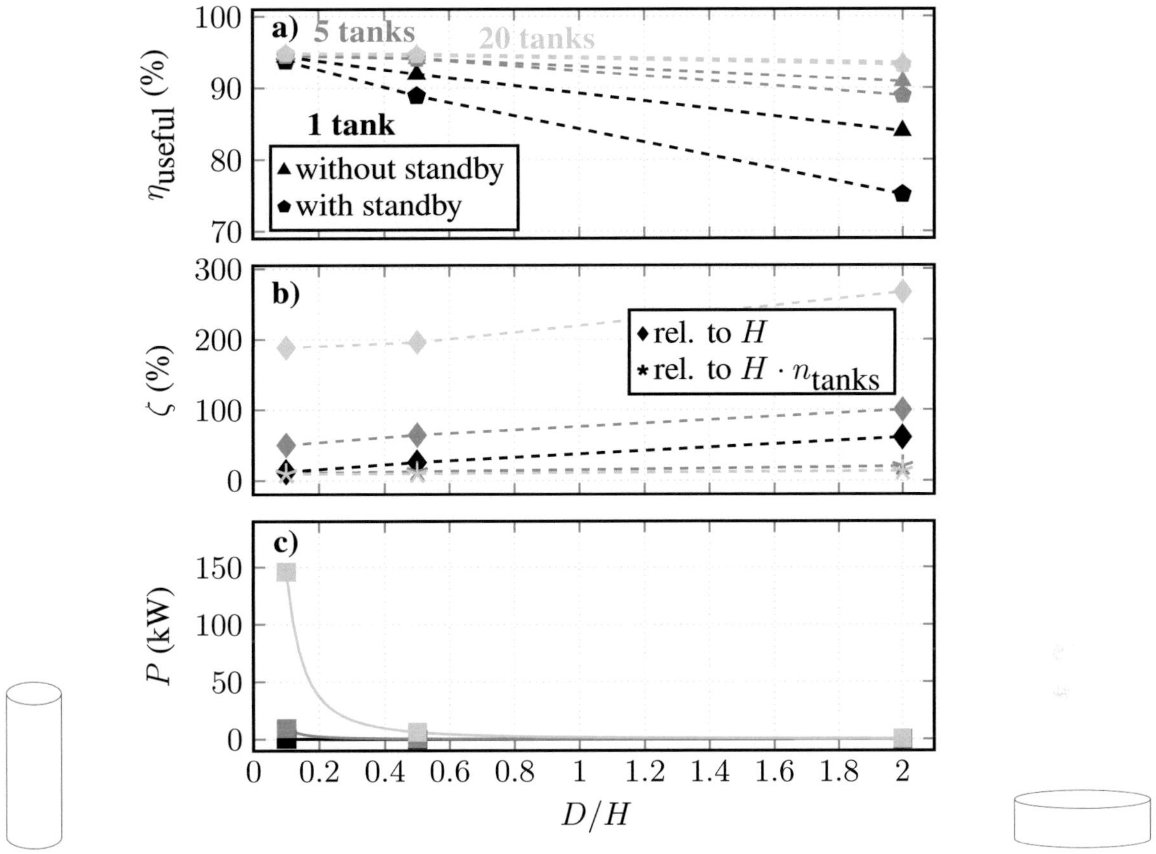

Figure 7.5: a) Discharge efficiency η_{useful} vs. D/H for 1 (black), 5 (grey) and 20 tanks (lightgrey); b) thermocline thickness after 8 h standby ζ vs. D/H; c) pumping power P vs. D/H; d) storage material cost M vs. D/H

has already been shown in Section 6.3.2. The greater the number of tanks, the weaker the influence of the D/H ratio, as the influence of thermal conduction in axial direction on the discharge efficiency is reduced with an increasing number of tanks owing to the thermal insulation between the tanks.

The standby period has a negligible effect on the discharge efficiency at low D/H ratios for all numbers of tanks investigated. For $D/H = 2.0$, a standby period during discharge leads to a reduction of the discharge efficiency in the single-tank arrangement: 83.9% (without standby) compared with 75.1% (with standby). For greater numbers of tanks the influence of standby is minimized. The thermocline degradation during standby increases with increasing D/H ratios for all numbers of tanks investigated. The thermocline thickness related to the overall tank height shows that the D/H ratio has a negligible influence on the thermocline degradation when more than five tanks are used.

The pumping power required to force the heat transfer fluid through the packed-bed storage is highest for small D/H ratios because of high fluid velocity and large tank heights. Further, an increased number of tanks leads to an increase of the pumping power. Thus, for 20 tanks and $D/H = 0.1$, highest pumping powers are calculated at $\approx$ 150 kW, which is $\approx$ 4% of an electric output of 4 MW, and thus no longer within a negligible range. The Ergun equation is valid at all parameters in this study according to Stieß [172] with a maximum Reynolds number of $Re_p = 2500$ for 20 tanks and $D/H = 0.1$.

All in all, highest efficiencies are reached at low D/H ratios for all numbers of tanks. At high D/H ratios the efficiency increases with increasing number of tanks, while at low D/H the efficiency is nearly independent of tank number. However, the greater the number of tanks (here maximum 20 tanks), the more critical is the resulting pumping power at those low D/H ratios. The influence of the D/H ratio on the efficiency decreases with increasing tank number.

7.3.3 Multi-chamber arrangement

As the D/H ratio shows a smaller influence on the discharge behaviour with a large number of tanks than with a single-tank arrangement, a multi-chamber arrangement is discussed in the following. This configuration could have the advantage of lower tank material and insulation costs compared with the multi-tank arrangements discussed before. The different chambers would have to be thermally insulated from each other and thus maintain a good stratification during standby. Additionally, the required pumping power can be reduced compared with that for a multi-tank with low D/H. This arrangement is proposed in the literature for hot-water storage tanks, e.g. in Ref. [186]. In this study, the arrangment consists of 20 serially charged and discharged packed-bed storage tanks with an even further increased D/H ratio ($D/H = 20$), allowing the tanks to be installed in a single tank in a multi-chamber arrangement. This is illustrated in Figure 7.6. The resulting height of each tank is 0.36 m with a diameter of 7.2 m.

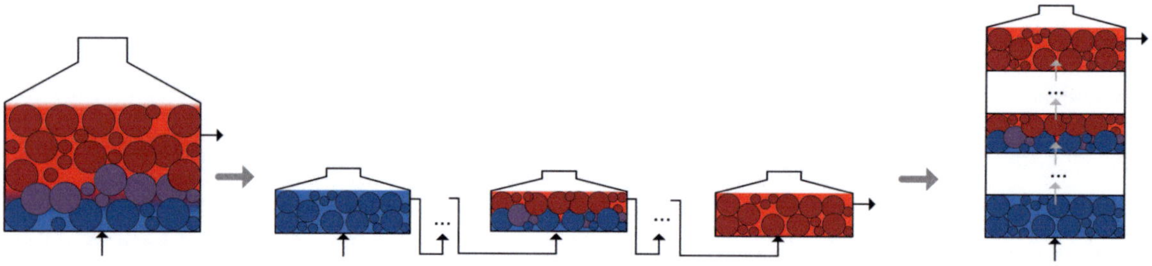

Figure 7.6: Scheme of multi-tank arrangement for high D/H ratios (during discharge) and scheme of multi-chamber arrangement

The resulting discharge efficiency is only 87.7%, which is $\approx 7\%$ lower than the arrangement with a single tank ($D/H = 0.1$) or five tanks ($D/H = 0.1$). It is approximately as efficient during discharge as the single-tank arrangement with the same dimensions ($D/H = 1.0$). After 8 h of standby, the thermocline zone occupies four of the 20 tanks/chambers, which is 20% of the overall tank height. The standby period has no significant influence on the discharge efficiency, which is reduced by only 0.5%. However, when 5 tanks are used with $D/H = 0.1$ the thermocline is only 10% of the overall tank height after 8 h of standby, and the efficiency is reduced by only 0.5%. The multi-chamber maintains its advantage during standby, because in the single tank the thermocline will eventually fill the whole tank, whereas the remaining tanks without the thermocline zone stay on temperature in the multi-chamber arrangement, when ideally insulated.

All in all, a multi-chamber arrangement, which is based on a high D/H ratio and a large number of tanks, is not advantageous from the view of discharge efficiency. A reduced number of tanks with a small D/H ratio will lead to higher discharge efficiencies. However, this configuration could still be of interest from an economic point of view, as only one tank with insulting layers is necessary, meaning reduced overall tank material and insulation costs. It might also be of interest if long standby periods are to be expected.

7.3.4 Gemasolar storage capacity

In this section, the applicability of a packed-bed storage with sodium on a state-of-the-art level is tested. The capacity of the two-tank storage system in the CSP plant Gemasolar is used with 1800 MWh with a storage time of 15 h. The two storage tanks have a height of 10.5 m, a diameter of 23 m and are filled with 8500 tons of solar salt. This capacity is now applied to single- and multi-tank systems with sodium (Figure 7.7).

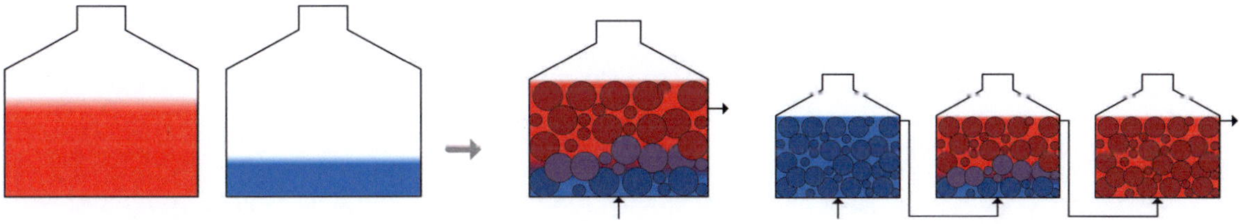

Figure 7.7: Scheme of two-tank arrangement installed in Gemasolar and the investigated corresponding single- and multi-tank arrangements

The packed-bed storage parameters of the reference case (Section 6.2) are used with sodium as heat transfer fluid. Exceptions are the storage capacity, discharge time and the temperatures of $T_{\mathrm{min}} = 290\,°\mathrm{C}$ and $T_{\mathrm{max}} = 565\,°\mathrm{C}$, which are adapted to the Gemasolar conditions. An efficient single-tank arrangement, with low D/H ratios, results in oversized storage tanks with tank heights of 107.9 m ($D/H = 0.1$) to 36.9 m ($D/H = 0.5$) (Table 7.4). A multi-tank arrangement with up to 20 tanks still leads to too large tank heights for $D/H = 0.1$; for 20 tanks a tank height of 39.7 m is necessary. For $D/H = 0.2$ the tank height for 20 tanks is 25.0 m and still large but in the range of a state-of-the-art storage tank height, which is ≈ 20 m [178].

The simulation of the packed-bed thermal storage is performed using an arrangement of one, three, five and seven tanks ($D/H = 0.5$). Furthermore, an arrangement with 20 tanks is simulated to allow a lower D/H ratio ($D/H = 0.2$). Additionally, a single-tank with a higher D/H ratio is investigated with a realistic tank height. The resulting discharge efficiency, standby behaviour and pumping power are presented in Table 7.5.

Table 7.4: Tank geometries for different numbers of tanks and D/H ratios for a storage size as that of Gemasolar power plant

D/H	0.1	0.2	0.5	1.0
n_{tanks}	1 .. 20	1 ... 20	1 ... 20	1 ... 20
H (m)	107.9 ... 39.7	68.0 ... 25.0	36.9 ... 13.6	23.2 ... 8.6
D (m)	10.8 ... 4.0	13.6 ... 5	18.5 ... 6.8	23.2 ... 8.6

Table 7.5: Tank heights and simulation results for different numbers of tanks and D/H ratios for a storage size as that of Gemasolar power plant; *related to overall tank height

D/H	0.2			0.5		1.0
n_{tanks}	20	1	3	5	7	1
H (m)	25.0	36.9	25.6	21.6	19.3	23.2
η_{useful} (%)	96.5	94.1	95.5	95.8	95.9	93.9
ζ after 8 h (%)	76.5	11.2	20.2	28.6	37.1	14.6
ζ^* after 8 h (%)	3.8	11.2	6.7	5.7	5.3	14.6
η_{useful} incl. 8 h standby (%)	96.5	93.4	95.2	95.7	95.8	92.3
P (kW)	717.0	0.5	3.3	8.6	16.1	0.2

The discharge efficiency η_{useful} is increased when a multi-tank arrangement is used compared with a single tank, while maintaining the same D/H ratio. Again, using more than five tanks shows no significant improvement in the discharge efficiency. It can be further improved in an arrangement with 20 tanks and an even lower D/H ratio. The thermocline degradation during standby is maximum in a single-tank arrangement, as the whole volume is affected. It is improved in a multi-tank arrangement with three and five tanks, but only slighty for more than five tanks. Again, if longer standby times are required, using more than five tanks can become more advantageous, as discussed previously. The results also show that in large-scale storage systems the discharge efficiency is not as sensitive regarding the D/H ratio as in smaller storage systems such as the one discussed before[7].

The increase of the required pumping power with a rising number of tanks has been described before. For the investigated multi-tank arrangement of 20 tanks, the pumping power is highest

[7] A more detailed analysis within the framework of the Master thesis of M. Meßmer (2018) has also shown that in larger storage systems with high discharge times, and thus low fluid velocities, the discharge efficiency is less sensitive regarding the D/H ratio and the filler diameter, but still regarding the porosity compared with a small-scale storage.

due to the high number of tanks and the low D/H ratio and is 3.6% of the electric output of Gemasolar (20 MW), which is no longer negligibe. The Ergun equation used is valid for all multi-tank arrangements according to Stieß [172] and, with exception of the arrangement with 20 tanks, also according to Allen et al. [170].

The cost of the storage materials used in the packed-bed system with sodium and quartzite as filler material ($\varepsilon = 0.22$) is 8.2 €/kWh for each arrangement considered in this study. For the solar salt quantity used in Gemasolar (8500 tons), material costs of 11.3 €/kWh are determined for the two-tank configuration. If solar salt were used in the packed-bed system investigated here, the material costs would decrease to 6.7 €/kWh; however, there would also be a decrease in the discharge efficiency compared with using sodium, as discussed in Section 6.5.1.

All in all, a packed-bed thermal energy storage with sodium can be realized with a storage size as implemented in Gemasolar. A five-tank arrangement shows higher efficiencies than a single tank, during both discharge and standby. Using more than five tanks leads to no significant improvement. The discharge efficiency of the currently installed two-tank solution is $\approx$ 100% for a well-insulated storage tank [94]. In the temperature range of Gemasolar the storage medium cost of a packed-bed storage system with sodium and quartzite is in the same range as the medium cost of the two-tank solar salt solution.

7.4 Conclusion

In order to realize low D/H ratios and at the same time keep a reasonable tank height, a multi-tank arrangement is proposed. Additionally, the serial arrangement of multiple tanks can reduce the thermocline degradation during standby, as only the tanks holding the thermocline are affected during standby; the remaining tanks stay at minimum or maximum temperature, assuming ideal insulation.

The number of tanks was varied at constant D/H ratio, which leads to the conclusion that the discharge efficiency and the standby behaviour can be improved by using more than one tank up to a certain number of tanks, which is five tanks for the reference case investigated. For a higher number of tanks the performance was not significantly enhanced. Additionally, the pumping power increased for an increasing number of tanks. Furthermore, the D/H ratio was varied for a single-, five- and twenty-tank arrangement. Highest discharge efficiencies and best standby behaviour were reached at low D/H ratios ($D/H = 0.1$). At those low ratios, the influence of the number of tanks on the discharge and standby performance was negligible. The required pumping power for a large number of tanks at low D/H ratios became no longer negligible. In conclusion, multi-tank arrangements enable low D/H ratios and high efficiencies; the single-tank height would exceed technical feasibily at that ratio, e.g. if the storage tank were installed indoors.

Because the influence of the D/H ratio on the discharge behaviour is not as significant for a high number of tanks as it is for a single-tank arrangement, a multi-chamber arrangement was investigated with horizontal, thermally insulting layers. This arrangement also requires less tank material and fewer insulation costs and less pumping power compared with the multi-tank arrangements. However, the results led to the conclusion that this arrangement can be profitable only if an economic analysis of the additional costs is included, as the discharge efficiency was decreased by 5% compared with a five-tank arrangement with a low D/H ratio and the same capacity.

The investigation of a multi-tank arrangement for a thermal energy storage system the size of the one installed in Gemasolar showed that the proposed storage system could be realized with sodium as heat transfer fluid for this capacity and temperature range. A discharge efficiency of $\approx 96\%$ was reached with an arrangement of five tanks. The results also showed that for sufficiently large capacities, and thus tank heights, the discharge efficiency is not as sensitive regarding the D/H ratio as for smaller capacities such as the selected reference case. The storage material costs were in the same range as the cost of solar salt used in the two-tank configuration.

8 Summary and outlook

8.1 Summary

In currently operational solar tower plants, solar salt is commonly used as heat transfer and storage fluid in a two-tank configuration. Liquid metals, especially sodium, are proposed as alternative fluids in the thermal receiver. Their high thermal conductivity makes it possible to build smaller tubular receivers, while maintaining a small temperature difference between wall and bulk, and thus reducing the heat losses and material costs. Significant reductions of the LCOE have been shown recently. Additionally, sodium can be used above the temperature limit of solar salt (600 °C), thus allowing the coupling to advanced power cycles, leading to higher overall plant efficiencies. However, for this temperature level, the state-of-the-art two-tank storage solution with solar salt is no longer applicable.

Up until now, no alternative solutions to the two-tank arrangement have been tested for liquid metals. Therefore, in the present work, a systematic evaluation of suitable thermal energy storage systems for sodium and temperatures beyond 600 °C was performed and a thermocline storage system with filler material was selected. In order to theoretically investigate the selected storage system, a two-dimensional two-phase model was developed, verified and validated to calculate the temperature distribution in the fluid and solid in the storage during cycling (charge/discharge) and standby, in order to assess the performance of sodium in packed-bed storage systems. The calculated temperature distributions in the tank were evaluated using the discharge efficiency η_{useful}, the relative height of the thermocline zone ζ, the required pumping power and the storage-material cost.

The parametric study showed that highest discharge efficiencies and best standby behaviour can be reached with a low porosity, large tank heights with small diameters and small filler particles. The pumping power increased when these parameters were decreased; however, the absolute values were still very low. The storage-material cost also decreased with low porosity owing to the large amount of low-cost solid filler material. Those low porosities can be attained only when multi-disperse packed beds are used; at the same time a small mean diameter must be maintained, as this is also beneficial.

The comparison with solar salt at conventional temperatures and alternative salt mixtures at elevated temperatures showed higher discharge efficiencies for sodium, $\approx 90\%$ compared with $\approx 87\%$.

This is because of the excellent heat transfer between fluid and filler particles. However, using sodium in a packed-bed storage led to an increased thermocline degradation during standby due to its high thermal diffusivity. A standby period of 8 h during discharge led to a discharge efficiency reduction of $\approx 4\%$ in the case of sodium and of only $\approx 0.5\%$ for salts. In conclusion, only short standby periods need to be realized when using sodium in a single-tank arrangement; otherwise, technological solutions need to be found to reduce the thermocline degradation during standby, such as using a multi-tank arrangement.

The investigation of serially connected tanks in a multi-tank arrangement showed that up to a certain number of tanks, the discharge efficiency increased, e.g. from $\approx 92\%$ (single tank) to $\approx 94\%$ (five tanks) for the reference case chosen in the present work. Additionally, the degradation of the thermocline during standby was reduced, as only the tanks containing the thermocline zone are affected, e.g. from $\approx 26\%$ (single-tank) to $\approx 13\%$ (five tanks) taking the overall tank height into account. The remaining ones stayed at charge or discharge temperature, when ideal insulation is assumed. For a higher number of tanks, the discharge efficiency and the standby behaviour improved only slightly, while the pumping power increased.

The multi-tank arrangement also makes it possible to realize lower D/H ratios and thus, increased discharge efficiency and less thermocline degradation during standby related to the overall tank height. A comparison of three different tank numbers with varying D/H ratios showed that for very low ratios of $D/H = 0.1$, a single-tank arrangement was as efficient as a multi-tank arrangement, even if an 8-h standby period was included, and the pumping power was also lower. However, if the space is limited or the tank dimensions are technically not feasible, a multi-tank arrangement is the efficient alternative. Furthermore, for longer standby periods, the multi-tank arrangement eventually becomes more efficient, as the degradation is limited to the tanks occupying the thermocline zone. In conclusion, the multi-tank system can have advantages over the single-tank arrangement during both cycling and standby, but it is dependent on each application. For a storage capacity such as the one implemented in Gemasolar, a multi-tank arrangement is necessary to reach low D/H ratios at technically feasible tank heights and led to discharge efficiencies of $\approx 96\%$, when five tanks were used with a D/H ratio of 0.5. It was also shown that the sodium storage would not be significantly more expensive than a solar salt system regarding material costs.

All in all, the preceding simulations showed that sodium can operate within the same range of performance parameters as state-of-the-art molten salt in a packed-bed storage. The results encourage further investigation into a sodium-based CSP system that also has a sodium-based storage system.

8.2 Outlook

Based on the findings presented here, the following experimental and theoretical research questions and subsequent work have been identified.

The selected storage system should be demonstrated with sodium as heat transfer fluid in a lab-scale experiment. In doing so, the following issues need to be addressed:

a) Isothermal and cyclic testing of candidate filler materials at high temperatures ($T > 600\,^{\circ}C$) in contact with sodium should be performed, as this has not yet been done to the best of the author's knowledge. A selection of filler materials is proposed in this work.
b) The axial temperature distribution of the fluid in a demonstration packed-bed storage tank should be determined in both cyclic and standby tests. With those results the assumptions made in the model regarding the effective thermal conductivity in the bed and the heat transfer between fluid and filler can be validated. This could also show the necessity for fundamental research work concerning the Nusselt correlation for the heat transfer of liquid metal to filler material in a packed bed, as no such correlation is available to date. Additionally, the potentially poor wetting of the filler material with sodium at low temperatures (e.g. when the tank is being filled) mentioned in this work should be tested.
c) The pumping power required to force the fluid through the packed bed should be determined, as references show that the Ergun equation used in this work can underestimate the pressure loss.

Furthermore, a detailed thermo-economic analysis (as done by Fritsch [7]) should be added, where the levelized costs of electricity resulting from the whole CSP plant with sodium as heat transfer fluid including the proposed thermal energy storage system are evaluated.

A further packed-bed configuration that could be simulated with the code presented in this work, applying few modifications, is the use of encapsulated phase-change material. The use of a packed-bed storage system with encapsulated phase-change material could lead to an increased efficiency and reduced storage costs, as already shown for molten salts in the literature.

A Appendix

A.1 Overview of installed and planned solar tower plants

A list of the operational and planned solar tower plants both with and without thermal energy storage system is given in Tables A.1 and A.2.

A.2 Energy-intensive industries with potential for heat recovery

In the reports of Fraunhofer ISI [70] and of the U.S. Department of Energy [69] the following industries with potential for heat recovery at elevated temperatures are listed: iron and steel, metal (non-iron), cement, glass, and chemical.

The iron and steel industry is with 4–5% one of the main primary energy consumers of the world [189]. Approximately 80% of the steel is produced in a blast furnace from coke and ore. The remaining 20% is produced in electric arc furnaces from scrap steel. In several steps in both routes heat is produced at high temperatures [70]. The challenge of using the high-temperature heat in the steps of the primary route is the composition of the high-temperature exhaust gases [69]. To clean them, they have to be cooled down [70]. Waste-heat boilers can be utilized to use the heat from the off-gases [69]. During production in the electric arc furnace, temperatures of up to 1900 °C can be reached in the exhaust gas stream, which are typically used to pre-heat the scrap metal [70].

The energy usage of the aluminium industry dominates the production of metals besides steel (copper, zinc, lead and noble metals). Similar to steel, aluminium can be produced from minerals or from scrap aluminium. The energy consumption via the primary route is six times higher than via the secondary route [69, 70]. If aluminium is produced from raw materials, electrolysis is used at temperatures of 960 °C. The main heat losses are at the electrolysis cell walls and the excess heat is not yet being re-used. If it is produced from scrap aluminium in a furnace at up to 1200 °C, the excess heat can be re-used only for pre-heating at low temperatures [70].

The cement production process consists of several steps including mining, crushing raw material, clinker production and milling [69, 70]. Among these, the clinker production is the most energy-

Table A.1: Installed and planned solar tower plants in alphabetical order (1/2)

Name	Country	Power (MW_e)	Start of operation	Heat transfer fluid	Storage time (h)	Ref.
Atacama-1	Chile	110	2018	Molten Salt	17.5	[22]
Atacama-2	Chile	110	2019	Molten Salt	15	[187]
Aurora	Australia	150	2020	Molten salt	8	[22]
Copiapo	Chile	260	2019	Molten Salt	14	[22]
Crescent Dunes	USA	110	2015	Molten Salt	10	[22]
Dahan Power Plant	China	1	2012	Water/steam	1	[22]
DEWA	UAE	100	2020	Molten salt	15	[22]
Gemasolar	Spain	20	2011	Molten salt	15	[22]
Delingha	China	135	2018	Molten Salt	3.7	[22]
Golden Tower	China	100	2020	Molten salt	8	[22]
Golmud	China	200	2018	Molten salt	15	[22]
Greenway Mersin	Turkey	1	2012	Water/steam	4	[22]
Hami	China	50	2019	Molten Salt	8	[22, 23]
Jemalong 1	Australia	1	2016	Liquid sodium	3	[22]
Jemalong 2	Australia	10	2019	Liquid sodium	4	[188]
Jülich	Germany	1.5	2008	Ceramic	1.5	[22]
Khi Solar One	South Africa	50	2016	Water/steam	2	[22]
Likana	Chile	390	2021	Molten Salt	13	[22]
MINOS	Greece	52	2020	Molten Salt	5	[22]

Table A.2: Installed and planned solar tower plants in alphabetical order (2/2)

Name	Country	Power (MW_e)	Start of operation	Heat transfer fluid	Storage time (h)	Ref.
Noor III	Marocco	150	2017	Molten Salt	7	[22]
PS 10	Spain	10	2007	Water/steam	1	[22]
PS 20	Spain	20	2009	Water/steam	1	[22]
Qinghai Gonghe	China	50	2020	Molten Salt	6	[22]
Qinghai Supcon	China	50	2018	Molten Salt	6	[22]
Redstone	South Africa	100	2018	Molten Salt	12	[22]
Shangyi	China	50	2019	Molten Salt	4	[22]
SunCan Dunhuang I	China	10	2009	Molten Salt	15	[22]
SunCan Dunhuang II	China	100	2018	Molten Salt	11	[22]
Tamarugal	Chile	450	2021	Molten Salt	13	[22]
Yumen 100	China	100	2030	Molten Salt	10	[22, 23]
Yumen 50	China	50	2018	Molten Salt	6	[22, 23]
ACME	India	2.5	2011	Water/steam	0	[22]
Ashalim	Israel	121	2017	Water/steam	0	[22]
Ivanpah	USA	392	2014	Water/steam	0	[22]
Sierra SunTower	USA	5	2009	Water/steam	0	[22]
Sundrop	Australia	1.5	2016	Water/steam	0	[22]

intensive. The excess heat in the rotary kiln and following cooling of the clinker bricks is used to pre-heat the raw materials and the combustion air. Furthermore, electricity is already produced from excess heat in four plants in the United States [69]. Still, a large amount of waste heat from kiln exhaust gases and clinker cooling is not re-used [69].

Glass manufacturing takes place at temperatures of up to 1550 °C. Waste heat from the furnaces is used to pre-heat the combustion air. Furthermore, cullet material is pre-heated and electricity is generated from waste-heat boilers [69, 70]. The US DOE report [69] states that, although heat recovery is already common practice, there is still potential for more implementation, especially in the regenerative furnaces.

In the chemical industry, the processes are complex, optimized and coupled to each other; therefore, heat integration is not easily realizable. Two processes are highlighted in the Fraunhofer ISI report [70]: ethylene and ammonia production. Ethylene is produced in a steam-cracking process. The raw materials are heated up to 650 °C and superheated steam is then added to crack them at 850 °C. The furnace includes a radiative part, where the reaction takes place, and a convection part, where heat is exhanged with process fluids [69]. Ammonia is produced in two steps: a synthesis gas process and ammonia synthesis. The energy-intensive part is the production of synthesis gas. The exhaust gases are used to produce steam, which is re-used in the process [70].

A.3 Physical properties

The correlations for the physical proporties of the investigated heat transfer fluids are listed in the following with the temperature T in K. The unit of density ρ is kg/m^3, dynamic viscosity μ in Pa · s, specific heat capacity c_p in J/(kgK) and thermal conductivity λ in W/(mK).

Sodium from Sodium-NaK engineering handbook [139]:

$$\begin{aligned}\rho = 950.1 - 0.22976 \cdot (T - 273.15) + 1.46 \cdot 10^{-5} \cdot (T - 273.15)^2 \\ + 5.638 \cdot 10^{-9} \cdot (T - 273.15)^3\end{aligned} \tag{A.1}$$

$$\mu = \begin{cases} 0.1235 \cdot 10^{-3} \cdot (\rho/1000)^{1/3} \cdot \exp(0.697 \cdot \rho/T) & T < 773.15\,\text{K} \\ 0.0851 \cdot 10^{-3} \cdot (\rho/1000)^{1/3} \cdot \exp(1.040 \cdot \rho/T) & T > 773.15\,\text{K} \end{cases} \tag{A.2}$$

$$\begin{aligned}c_p = (0.34324 - 1.3868 \cdot 10^{-4} \cdot (T - 273.15) \\ + 1.1044 \cdot 10^{-7} \cdot (T - 273.15)^2) \cdot 4.184 \cdot 10^3\end{aligned} \tag{A.3}$$

$$\lambda = 91.8 - 4.9 \cdot 10^{-2} \cdot (T - 273.15) \tag{A.4}$$

Solar salt: $NaNO_3$-KNO_3 (composition by weight 0.60-0.40) from Zavaico [190].

$$\rho = 2090 - 0.636 \cdot (T - 273.15) \tag{A.5}$$

$$\mu = (22.714 - 0.12 \cdot (T - 273.15) + 2.281 \cdot 10^{-4} \cdot (T - 273.15)^2 - 1.474 \cdot 10^{-7} \cdot (T - 273.15)^3) \cdot 10^{-3} \tag{A.6}$$

$$c_p = 1443 + 0.172 \cdot (T - 273.15) \tag{A.7}$$

$$\lambda = 0.443 + 1.9 \cdot 10^{-4} \cdot (T - 273.15) \tag{A.8}$$

LBE from Handbook on Lead-bismuth Eutectic Alloy and Lead Properties, Materials Compatibility, Thermal-hydraulics and Technologies [191].

$$\rho = 11096 - 1.3236 \cdot T \tag{A.9}$$

$$\mu = 0.494 \cdot 10^{-3} \cdot \exp(754.1/T) \tag{A.10}$$

$$c_p = 159 - 2.72 \cdot 10^{-2} \cdot T + 7.12 \cdot 10^{-6} \cdot T^2 \tag{A.11}$$

$$\lambda = 3.61 + 1.517 \cdot 10^{-2} \cdot T - 1.741 \cdot 10^{-6} \cdot T^2 \tag{A.12}$$

Lead from Handbook on Lead-bismuth Eutectic Alloy and Lead Properties, Materials Compatibility, Thermal-hydraulics and Technologies [192].

$$\rho = 11441 - 1.2795 \cdot T \tag{A.13}$$

$$\mu = 4.55 \cdot 10^{-4} \exp(1069/T) \tag{A.14}$$

$$c_p = 175.1 - 4.961 \cdot 10^{-2} \cdot T + 1.985 \cdot 10^{-5} \cdot T^2 - 2.099 \cdot 10^{-9} \cdot T^3 - 1.524 \cdot 10^{6} \cdot T^{-2} \tag{A.15}$$

$$\lambda = 9.2 + 0.011 \cdot T \tag{A.16}$$

HTS 1: $ZnCl_2$-NaCl-KCl (composition by weight 0.686-0.075-0.239) from Li et al. [193].

$$\rho = 2.878 \cdot 10^{3} - 0.926 \cdot T \tag{A.17}$$

$$\mu = 0.121 \cdot \exp(-T/204.709) + 4.976 \cdot 10^{5} \cdot \exp(-T/29.917) + 3.41 \cdot 10^{-3} \tag{A.18}$$

$$c_p = 900 \tag{A.19}$$

$$\lambda = 0.514 - 2.331 \cdot 10^{-4} \cdot T \tag{A.20}$$

HTS 2: $MgCl_2$-KCl (composition by weight 0.375-0.625) from Mehos et al. [4], Li et al. [194] and Williams [195].

$$\rho = 1660 \tag{A.21}$$

$$\mu = 5 \cdot 10^{-3} \tag{A.22}$$

$$c_p = 1150 \tag{A.23}$$

$$\lambda = 0.4 \tag{A.24}$$

HTS 3: Na_2CO_3-K_2CO_3-Li_2CO_3 (composition by weight 0.334-0.345-0.321) from An et al. [196].

$$\rho = (2.27 - 4.3410^{-4} \cdot T) \cdot 1000 \tag{A.25}$$

$$\mu = 169.8 \cdot \exp(-0.013 \cdot T) + 0.265 \cdot \exp(-0.004 \cdot T) \tag{A.26}$$

$$c_p = 1612 \tag{A.27}$$

$$\lambda = 0.469 \tag{A.28}$$

The physical properties of the investigated filler materials are listed in Table A.3. They are assumed to be temperature-independent in the temperature range considered.

Table A.3: Candidate filler materials and their physical properties at *0 °C, $^+$400 °C, ^{1}20 °C

Material	Comp.	ρ (kg m^{-3})	c_p (J kg^{-1}K^{-1})	λ (Wm^{-1}K^{-1})	Ref.
Quartzite	SiO_2	2640	1050	2.5	[87]
Spinel	$MgAl_2O_4$	2850	1050^+	3.8^+	[135]
Corundum	Al_2O_3	3200	1011^+	5^+	[135]
Aust. steel	X 5 CrNi 18-10	7900^1	560^+	21^+	[135]
Iron	Fe	7870*	603^+	84*	[135]

A.4 Cost data

The specific material cost data of the heat transfer fluids investigated for use in packed bed thermal energy storage systems in this work are listed in Tables A.4 (liquid metals) and A.5 (molten salts).

The specific material cost data of selected solid filler materials are shown in Table A.6.

Table A.4: Specific material cost data of liquid metals, LBE = lead bismuth eutectic

	Sodium	LBE	Lead
C (€/kg)	2.6	12	1.6
Ref.	[44]	[44]	[44]

Table A.5: Specific material cost data of molten salts, solar salt := $NaNO_3$-KNO_3 (0.60-0.40), HTS 1 := $ZnCl_2$-NaCl-KCl (0.686-0.075-0.239), HTS 2 := $MgCl_2$-KCl (0.375-0.625) and HTS 3 := Na_2CO_3-K_2CO_3-Li_2CO_3 (0.334-0.345-0.321)

	Solar salt	HTS 1	HTS 2	HTS 3
C (€/kg)	1.0	1.3	0.4	2.6
Ref.	[44, 197]	[198, 199]	[199, 200]	[199, 201]

A.5 Modelling of a packed bed storage with sodium – cycling

A.5.1 Literature review

In Table A.7, the models proposed in the literature are arranged by accuracy from the one-phase (1P) one-dimensional (1D) model, where solid and fluid phase are considered as one homogeneous phase, to the two-phase (2P) two-dimensional (2D) model including intra-particle (p) diffusion. The subjects of the publications are added to illustrate the purposes each model might be suitable for.

Furthermore, computational fluid dynamics (CFD) models are used in the literature, mainly with the purpose of modelling thermo-mechanical stresses [101] and reproducing specific inlet shapes/conditions from the experiment [110, 204]. In these publications, the solid particles are not modelled by designing a suitable mesh, but by assuming an isotropic porous structure and using the Darcy-Forchheimer equation. If radiation plays a role, e.g. in the case of gas-solid regenerator type storages, a detailed mesh is used [205, 206].

For the goal envisaged in this work, which is to assess the performance of the heat transfer fluid sodium in a packed bed thermocline storage system, a CFD model is not efficient. Instead, efficient runs of numerous cases with varying parameters are key here, so that specific models are preferred, as follows.

As the filler and liquid sodium do not necessarily have similar thermal diffusivities and, as it should also be possible to test small tanks (where a 1P model has shown its shortcomings in the literature), a two-phase model is required. To be able to investigate effects on radial temperature distributions in the tank (e.g. due to heat losses) and to ensure a very general applicability of the

Table A.6: Specific material cost data of candidate filler materials

	Quartzite SiO_2	Spinel $MgAl_2O_4$	Corundum 85 % Al_2O_3	Aust. steel X 5 CrNi 18-10	Iron Fe
C (€/kg)	0.5	0.8–1.0	1.0	0.4–1.0	0.5–0.9
Ref.	[44]	[202]	[44, 203]	[202]	[202]

Table A.7: Models used in the literature, P = phase, D = dimension, p = particle

Model	General simplification	Investigations	Ref.
1P-1D	• fluid and solid considered as one phase • intra-particle gradient negligible: Bi < 0.1 • continuous solid phase • uniform porosity • radially uniform velocity • radially uniform temperature • fluid and solid similar thermal diffusivity	• introducing a dimensionless velocity and developing analytical design equation • comparison with 2P-model showing suitability for industrial tank sizes but not for small tanks	[149],[207]
2P-1D (Schumann)	• intra-particle gradient negligible: Bi < 0.1 • continuous solid phase • uniform porosity • radially uniform velocity • radially uniform temperature • diffusion terms negligible	• recomputing results from lab/pilot-scale experiments • parametric studies: tank and filler dimensions, porosity • comparison of fluids • different filler structures (Nu-numbers)	[39],[87], [88],[106], [208],[209]
2P-1D (Enhanced Schumann)	• intra-particle gradient negligible: Bi < 0.1 • continuous solid phase • uniform porosity • radially uniform velocity • radially uniform temperature	• recomputing results from lab/pilot-scale experiments • parametric studies: tank and filler dimensions • comparison of fluids • comparison with 1P-model showing suitability of 2P-model for all tank sizes • development of performance criteria	[91],[92], [93][110], [113],[117], [149],[210]
2P-1D (intra-p)	• uniform fluid temperature around solid particles • uniform porosity • radially uniform velocity • heat losses negligible (radially uniform temperature)	• ideal tank dimensions including PCM capsules	[167],[211]
2P-2D	• intra-particle gradient negligible: Bi < 0.1 • continuous solid phase • uniform porosity • radially uniform velocity	• thermo-mechanical studies (thermal ratcheting) • parametric studies: tank and filler dimensions, porosity • standby (including heat loss)	[89],[151]
2P-2D (intra-p)	• uniform fluid temperature around solid particles • uniform porosity • radially uniform velocity	• parametric studies: filler dimensions and material (exceeding Bi = 0.1)	[90]

code, a two-dimensional model is selected. Nevertheless, in most cases (e.g. parametric studies), assuming a well-insulated tank, a two- phase one-dimensional (2P-1D) model is sufficient.

In case of conventional heat transfer fluids used in the literature, namely molten salts and thermal oil, the Biot number is sufficiently low (Bi < 1) and thus, a continuous solid phase (lumped capacitance) can be assumed [156]. For example, a Biot-number of 0.1 means that the resistance on the outer surface is 10 times higher than the inner heat transfer resistance. This leads to a uniform temperature change along the radial axis of the filler spheres. Therefore, the heat transfer on the outside will be the limiting effect. For liquid sodium, however, the following estimation shows that the Biot number is significantly higher.

The Biot-number for a fluid in contact with a spherical particle is defined according to Equation A.29. The convection heat transfer coefficient α for the fluid flowing through the tank is dependent on the Nusselt number in the packed bed.

$$\mathrm{Bi} = \frac{\alpha \cdot d/2}{\lambda_\mathrm{s}} \qquad \alpha = \frac{\mathrm{Nu}_\mathrm{bed} \cdot \lambda_\mathrm{f}}{d} \tag{A.29}$$

Owing to low fluid velocities in the packed bed, the minimum Nusselt number $\mathrm{Nu}_\mathrm{bed} = 2$ (only conduction) can be taken as a conservative estimation. Then, the definition of the Biot number can be simplified to $\mathrm{Bi} = \lambda_\mathrm{f}/\lambda_\mathrm{s}$. With a thermal conductivity of an exemplary solid particle (quartzite) of $\lambda_\mathrm{s} = 2.5\,\mathrm{W/(mK)}$, the Biot number is 0.2 for solar salt ($\lambda_\mathrm{f} = 0.5\,\mathrm{W/(mK)}$ at 400 °C) and 28.9 for sodium ($\lambda_\mathrm{f} = 72.2\,\mathrm{W/(mK)}$ at 400 °C).

Owing to the high Biot-number ($\mathrm{Bi} \gg 1$) when liquid sodium is used, the thermal diffusion in the filler particle needs to be considered. Therefore, the 2D-2P model including intra-particle diffusion is selected for the determination of the temperature distribution in a packed-bed thermocline storage system with sodium as heat transfer fluid. For salt, a simplified 2D-2P model assuming lumped capacitance can be used.

A.5.2 Natural convection

In the literature, natural convection is usually neglected, as opposed to forced convection in thermocline packed-bed storage systems. According to Yang et al. [107], the solid material limits natural convection in those systems. In this section, the local natural convection at the spherical particles and at the tank wall is assessed.

According to Ref. [212] the Nusselt number for natural convection around a sphere is determined with the Rayleigh and Prandtl number as follows.

$$\mathrm{Nu} = 2 + 0.56\left[\left(\frac{\mathrm{Pr}}{0.846+\mathrm{Pr}}\right)\mathrm{Ra}\right]^{1/4} \tag{A.30}$$

The Rayleigh number is determined from the Grashof and Prandtl number with the thermal expansion coefficient β, gravity g, temperature difference between bulk and wall ΔT, the characteristic length L_c, the kinematic viscosity ν and the thermal diffusivity a.

$$\mathrm{Ra} = \mathrm{Gr} \cdot \mathrm{Pr} = \frac{\beta g \Delta T L_c^3}{\nu a} \tag{A.31}$$

Natural convection only plays a role if a temperature difference is present. Therefore, natural convection is estimated in the thermocline region, where fluid comes in contact with filler at a different temperature and a tank wall having a different temperature. The effects at the inlet, where cold fluid enters when charging and hot fluid when discharging, are not considered here. The temperature difference between fluid and filler is $\approx 0.5\,\mathrm{K}$ in the thermocline region from the simulation results in this work for the reference case. It is calculated from the fluid temperature and the temperature at the sphere surface from the previous time step:

$$\Delta T = T(i,j) - T_{\mathrm{p}}(i,j-1) \tag{A.32}$$

Between fluid and tank wall the same temperature difference in assumed. The characteristic length L_c is either a sphere diameter or a section of a tank wall of the same length between two spheres, as illustrated in Figure A.1b.

Figure A.1: a) Boundary layer flow around a single sphere (adapted from Ref. [213]); b) schematic illustration of the flow length at the tank wall in a packed bed

With the physical properties of sodium at 500 °C, a characteristic length of 15 mm and the thermal expansion coefficient $\beta = 2.418 \cdot 10^{-4}$/°C [139], the resulting Rayleigh number is 213 and the Nusselt number is 2.6 (for $\Delta T = 10\,\mathrm{K}$, Ra = $4.3 \cdot 10^3$ and Nu = 3.2). As shown in Section 5.1.3, increasing the heat transfer Nusselt number between fluid and filler from 2 to 10 does not lead to significant changes in the temperature distribution during discharge in the selected reference case. Thus, the influence of natural convection is neglected in this work.

During standby fluid and solid material (filler particles/ tank wall) are at the same temperature at each axial position in the tank. Therefore, the natural convection is neglected also when modelling the standby behaviour.

A.5.3 Axial thermal conduction in the solid phase

The axial heat conduction in the solid phase, assuming a lumped capacitance (Bi < 1), is mostly neglected in the literature. In order to proof that it is also negligible in the current work, the following comparison is made. The storage tank of Sandia National Laboratories with solar salt as heat transfer fluid [39] is taken as a reference case. The whole tank is at maximum temperature at the beginning and is discharged for one hour. The simulation is performed by including the axial conduction in the solid phase (Equation 5.4) and without the axial conduction, and thus using only the analytical solution (Equation 5.5) as shown in Section 5.1.1. In both cases the heat losses are set to zero. Figure A.2 shows that there is no significant difference between the results with and without including axial heat conduction of the solid phase included in the simulation. The curvature is slightly sharper with $k_{sx} = 0$, as expected, and the temperature of the fluid phase is always slightly below that of the solid phase, as it is a discharge process. All in all, this comparison shows that the axial heat conduction in the solid phase can be neglected and the analytical solution is sufficient.

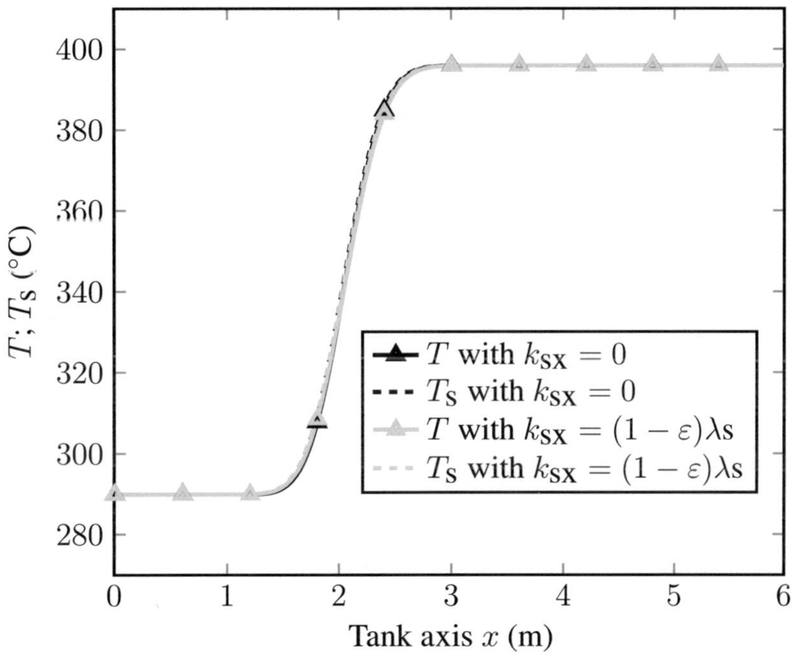

Figure A.2: Temperature in fluid (T) and solid (T_s) phase after 1 h discharge both with and without axial heat conduction included in the solid phase

A.5.4 Initial and boundary conditions

At the beginning of the first cycle, the initial condition (t = 0) is a uniform temperature in the tank. This temperature is T_{max} (fully charged) if the first cycling step is a discharge process. For charging, the initial temperature is T_{min} (fully discharged). Furthermore, a temperature distribution in the fluid, solid or particle can be applied (stratified temperature).

$$t=0: \quad \begin{cases} T(x,r) = T_\text{s}(x,r) = T_\text{p}(x,r,y) = T_\text{max} \\ T(x,r) = T_\text{s}(x,r) = T_\text{p}(x,r,y) = T_\text{min} \\ T(x,r) = f(x,r), T_\text{s}(x,r) = g(x,r), T_\text{p}(x,r,y) = h(x,r,y) \end{cases} \tag{A.33}$$

In Figure A.3 the boundary conditions for fluid, solid (homogeneous) and particle are displayed. The storage tank is discharged from the bottom and charged from the top. The axial coordinate x always starts from the inlet in the tank.

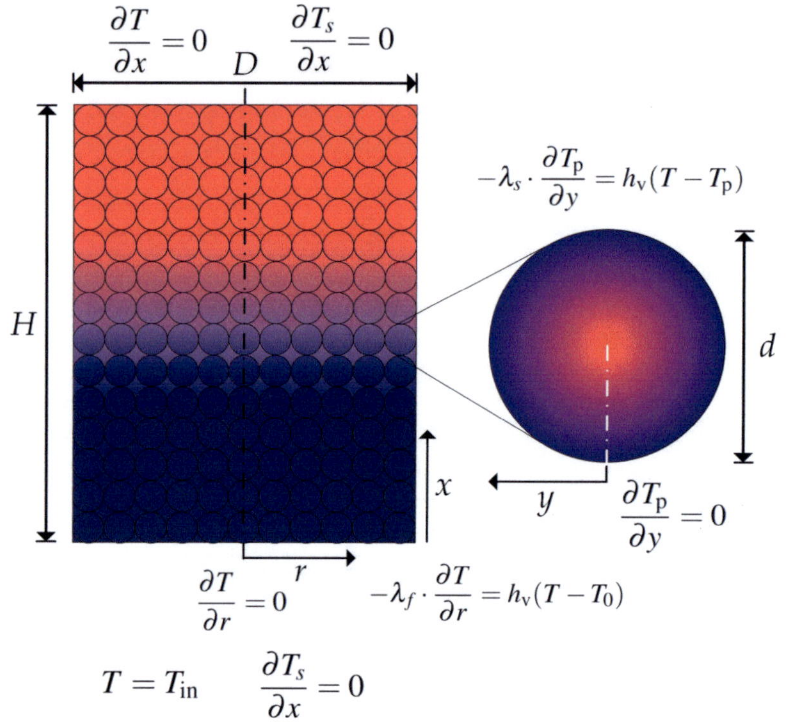

Figure A.3: Scheme of packed bed for the discharge process with the applied boundary conditions (cold fluid enters at the bottom of the tank)

At the inlet of the storage tank ($x = 0$), a boundary condition of the first kind with a constant inlet temperature T_{in} is imposed on the fluid term. In a charging process this temperature is $T_{\text{in}} = T_{\text{max}}$, in a discharging process $T_{\text{in}} = T_{\text{min}}$. At the outlet ($x = H$), a zero temperature gradient is assumed for the fluid. The solid material does not exchange heat with other solid material either at the inlet or at the outlet of the tank.

$$x=0: \qquad T = T_\text{in} \qquad \frac{\partial T_\text{s}}{\partial x} = 0 \tag{A.34}$$

$$x=H: \qquad \frac{\partial T}{\partial x} = 0 \qquad \frac{\partial T_\text{s}}{\partial x} = 0 \tag{A.35}$$

In the centre of the tank (r = 0), a symmetry boundary condition is applied for both fluid and solid. At the tank wall (r = R), a convection boundary condition can be imposed in case of two-dimensional problems, e.g. if the tank is not well-insulated and heat losses have to be considered. For the simplified 2P-1D model, an adiabatic boundary condition is applied at the tank wall. It is assumed that the heat transfer between tank wall and the spheres is negligible compared with the one between wall and fluid, owing to point contact of the spheres and a considerably lower thermal conductivity of the solid material than of the fluid.

$$r=0: \qquad \frac{\partial T}{\partial r}=0 \qquad \frac{\partial T_{\mathrm{s}}}{\partial r}=0 \tag{A.36}$$

$$r=R: \qquad -\lambda_{\mathrm{f}}\cdot\frac{\partial T}{\partial r}=h_{\mathrm{w}}(T-T_0) \qquad \frac{\partial T_{\mathrm{s}}}{\partial r}=0 \tag{A.37}$$

For the intra-particle model, the boundary conditions of the filler particles need to be set. In the core of the solid particles (y = 0), no heat is exchanged owing to the symmetry condition. At the surface ($y = d/2$), the convection boundary condition takes into account the heat transfer between fluid and filler material and links the fluid energy equation to the energy equation of the particle.

$$y=0: \qquad \frac{\partial T_{\mathrm{p}}}{\partial y}=0 \tag{A.38}$$

$$y=d/2: \qquad -\lambda_{\mathrm{s}}\cdot\frac{\partial T_{\mathrm{p}}}{\partial y}=\alpha(T-T_{\mathrm{p}}) \tag{A.39}$$

A.5.5 Finite volume method

The energy equation for the fluid (Equation 5.1/5.3) is integrated at each node both over the control volume and over time:

$$\begin{aligned}&\int_0^{2\pi}\int_{x_{\mathrm{s}}}^{x_{\mathrm{n}}}\int_{r_{\mathrm{w}}}^{r_{\mathrm{e}}}\int_t^{t+\Delta t}\left[\varepsilon\rho_{\mathrm{f}}c_{p\mathrm{f}}\frac{\partial T}{\partial t}+u\varepsilon\rho_{\mathrm{f}}c_{p\mathrm{f}}\frac{\partial T}{\partial x}\right]dt\,r dr\,dx\,d\varphi\\&=\int_0^{2\pi}\int_{x_{\mathrm{s}}}^{x_{\mathrm{n}}}\int_{r_{\mathrm{w}}}^{r_{\mathrm{e}}}\int_t^{t+\Delta t}\left[\varepsilon\lambda_{\mathrm{f}}\frac{\partial^2 T}{\partial x^2}+\varepsilon\lambda_{\mathrm{f}}\left(\frac{\partial^2 T}{\partial r^2}+\frac{1}{r}\frac{\partial T}{\partial r}\right)-h_{\mathrm{v}}(T-T_{\mathrm{s}})\right]dt\,r dr\,dx\,d\varphi\end{aligned} \tag{A.40}$$

The coefficients are shortened to C_1–C_5 for simplification.

$$\begin{aligned}&\int_0^{2\pi}\int_{x_{\mathrm{s}}}^{x_{\mathrm{n}}}\int_{r_{\mathrm{w}}}^{r_{\mathrm{e}}}\int_t^{t+\Delta t}\left[C_1\frac{\partial T}{\partial t}+C_2\frac{\partial T}{\partial x}\right]dt\,r dr\,dx\,d\varphi\\&=\int_0^{2\pi}\int_{x_{\mathrm{s}}}^{x_{\mathrm{n}}}\int_{r_{\mathrm{w}}}^{r_{\mathrm{e}}}\int_t^{t+\Delta t}\left[C_3\frac{\partial^2 T}{\partial x^2}+C_4\left(\frac{\partial^2 T}{\partial r^2}+\frac{1}{r}\frac{\partial T}{\partial r}\right)-C_5(T-T_{\mathrm{s}})\right]dt\,r dr\,dx\,d\varphi\end{aligned} \tag{A.41}$$

Each term is integrated separately (Equations A.42–A.46). In the following, each term of the fluid energy equation is solved separately in detail:

$$\int_0^{2\pi}\int_{x_{\mathrm{s}}}^{x_{\mathrm{n}}}\int_{r_{\mathrm{w}}}^{r_{\mathrm{e}}}\int_t^{t+\Delta t} C_1\frac{\partial T}{\partial t}dt\,r dr\,dx\,d\varphi\approx C_1\, r_{\mathrm{P}}\Delta r\,\Delta x\,2\pi(T_{\mathrm{P}}^{j+1}-T_{\mathrm{P}}^{j}) \tag{A.42}$$

$$
\begin{aligned}
&\int_0^{2\pi}\int_{x_s}^{x_n}\int_{r_w}^{r_e}\int_t^{t+\Delta t} C_2\frac{\partial T}{\partial x}dt\,rdrdxd\varphi \approx C_2 r_P \Delta r 2\pi \int_t^{t+\Delta t}(T_n - T_s)dt \\
&\approx C_2 r_P \Delta r 2\pi \int_t^{t+\Delta t}(\beta_n T_N - \beta_s T_S + T_P(\beta_s - \beta_n))dt \\
&\approx C_2 r_P \Delta r 2\pi \Delta t \left[\Theta(\beta_n T_N - \beta_s T_S + T_P(\beta_s - \beta_n))^{j+1}\right. \\
&\left. + (1-\Theta)(\beta_n T_N - \beta_s T_S + T_P(\beta_s - \beta_n))^{j}\right]
\end{aligned}
\tag{A.43}
$$

$$
\begin{aligned}
&\int_0^{2\pi}\int_{x_s}^{x_n}\int_{r_w}^{r_e}\int_t^{t+\Delta t} C_3\frac{\partial^2 T}{\partial x^2}dt\,rdrdxd\varphi \\
&= \int_0^{2\pi}\int_{x_s}^{x_n}\int_{r_w}^{r_e}\int_t^{t+\Delta t} C_3\frac{\partial}{\partial x}\left(\frac{\partial T}{\partial x}\right)dt\,rdrdxd\varphi \\
&\approx C_3 r_P \Delta r 2\pi \int_t^{t+\Delta t}\left[\left(\frac{\partial T}{\partial x}\right)_n - \left(\frac{\partial T}{\partial x}\right)_s\right]dt \\
&\approx C_3 r_P \Delta r 2\pi \int_t^{t+\Delta t}\left[\frac{T_N - T_P}{\Delta x_{PN}} - \frac{T_P - T_S}{\Delta x_{SP}}\right]dt \\
&\approx C_3 r_P \Delta r 2\pi \int_t^{t+\Delta t}\left[\frac{T_N}{\Delta x_{PN}} + \frac{T_S}{\Delta x_{SP}} - T_P\left(\frac{1}{\Delta x_{PN}} + \frac{1}{\Delta x_{SP}}\right)\right]dt \\
&\approx C_3 r_P \Delta r 2\pi \Delta t \left[\Theta\left(\frac{T_N}{\Delta x_{PN}} + \frac{T_S}{\Delta x_{SP}} - T_P\left(\frac{1}{\Delta x_{PN}} + \frac{1}{\Delta x_{SP}}\right)\right)^{j+1}\right. \\
&\left. + (1-\Theta)\left(\frac{T_N}{\Delta x_{PN}} + \frac{T_S}{\Delta x_{SP}} - T_P\left(\frac{1}{\Delta x_{PN}} + \frac{1}{\Delta x_{SP}}\right)\right)^{j}\right]
\end{aligned}
\tag{A.44}
$$

$$
\begin{aligned}
&\int_0^{2\pi}\int_{x_s}^{x_n}\int_{r_w}^{r_e}\int_t^{t+\Delta t} C_4\left[\frac{\partial^2 T}{\partial r^2} + \frac{1}{r}\frac{\partial T}{\partial r}\right]dt\,rdrdxd\varphi \\
&= \int_0^{2\pi}\int_{x_s}^{x_n}\int_{r_w}^{r_e}\int_t^{t+\Delta t} C_4\left[\frac{\partial}{\partial r}\left(r\frac{\partial T}{\partial r}\right)\right]dt\,drdxd\varphi \\
&\approx C_4 \Delta x 2\pi \int_t^{t+\Delta t}\left[r_e\left(\frac{\partial T}{\partial r}\right)_e - r_w\left(\frac{\partial T}{\partial r}\right)_w\right]dt \\
&\approx C_4 \Delta x 2\pi \int_t^{t+\Delta t}\left[r_e\frac{T_E - T_P}{\Delta r_{PE}} - r_w\frac{T_P - T_W}{\Delta r_{WP}}\right]dt \\
&\approx C_4 \Delta x 2\pi \Delta t \left[\Theta\left(r_e\frac{T_E}{\Delta r_{PE}} + r_w\frac{T_W}{\Delta x_{WP}} - T_P\left(\frac{r_e}{\Delta r_{PE}} + \frac{r_w}{\Delta r_{WP}}\right)\right)^{j+1}\right. \\
&\left. + (1-\Theta)\left(r_e\frac{T_E}{\Delta r_{PE}} + r_w\frac{T_W}{\Delta x_{WP}} - T_P\left(\frac{r_e}{\Delta r_{PE}} + \frac{r_w}{\Delta r_{WP}}\right)\right)^{j}\right]
\end{aligned}
\tag{A.45}
$$

$$\begin{aligned}
&\int_0^{2\pi}\int_{x_s}^{x_n}\int_{r_w}^{r_e}\int_t^{t+\Delta t} C_5(T-T_s)dt\,rdr\,dx\,d\varphi \\
&\approx C_5 r_P \Delta r \Delta x 2\pi \int_t^{t+\Delta t} T dt - C_5 r_P \Delta r \Delta x 2\pi \Delta t\, T_{s,P}^j \\
&\approx C_5 r_P \Delta r \Delta x 2\pi \Delta t \left[\Theta T_P^{j+1} + (1-\Theta) T_P^j\right] - C_5 r_P \Delta r \Delta x 2\pi \Delta t\, T_{s,P}^j
\end{aligned} \tag{A.46}$$

Inserting them back into Equation A.40 and dividing by $2\pi\Delta t$ leads to:

$$\begin{aligned}
&C_1 \frac{r_P \Delta r \Delta x}{\Delta t}(T_P^{j+1} - T_P^j) \\
&+ C_2 r_P \Delta r \left[\Theta(\beta_n T_N - \beta_s T_S + T_P(\beta_s - \beta_n))^{j+1}\right. \\
&+ \left.(1-\Theta)(\beta_n T_N - \beta_s T_S + T_P(\beta_s - \beta_n))^j\right] \\
&= C_3 r_P \Delta r \left[\Theta\left(\frac{T_N}{\Delta x_{PN}} + \frac{T_S}{\Delta x_{SP}} - T_P\left(\frac{1}{\Delta x_{PN}} + \frac{1}{\Delta x_{SP}}\right)\right)^{j+1}\right. \\
&+ \left.(1-\Theta)\left(\frac{T_N}{\Delta x_{PN}} + \frac{T_S}{\Delta x_{SP}} - T_P\left(\frac{1}{\Delta x_{PN}} + \frac{1}{\Delta x_{SP}}\right)\right)^j\right] \\
&+ C_4 \Delta x \left[\Theta\left(r_e\frac{T_E}{\Delta r_{PE}} + r_w\frac{T_W}{\Delta x_{WP}} - T_P\left(\frac{r_e}{\Delta r_{PE}} + \frac{r_w}{\Delta r_{WP}}\right)\right)^{j+1}\right. \\
&+ \left.(1-\Theta)\left(r_e\frac{T_E}{\Delta r_{PE}} + r_w\frac{T_W}{\Delta x_{WP}} - T_P\left(\frac{r_e}{\Delta r_{PE}} + \frac{r_w}{\Delta r_{WP}}\right)\right)^j\right] \\
&- C_5 r_P \Delta r \Delta x \left[\Theta T_P^{j+1} + (1-\Theta)T_P^j\right] + C_5 r_P \Delta r \Delta x\, T_{s,P}^j
\end{aligned} \tag{A.47}$$

The linear interpolation from the neighbouring nodes is performed using interpolation factors $\beta_n = \frac{x_n - x_P}{x_N - x_P}$ and $\beta_s = \frac{x_s - x_P}{x_S - x_P}$.

$$\begin{aligned}
T_n &= \beta_n T_N + (1-\beta_n) T_P \\
T_s &= \beta_s T_S + (1-\beta_s) T_P
\end{aligned} \tag{A.48}$$

In this work, a first-type upwind differencing scheme (UDS) is used for the spatial discretization of the advection terms: $\beta_n = 1$ and $\beta_s = 0$. For the diffusive terms, a central differencing scheme is used. For detailed information refer to Ref. [163].

The time scheme factor Θ allows selection of the time scheme:

$$\Theta = \begin{cases} 0 & \text{Explicit forward time scheme (FT)} \\ 0.5 & \text{Mixed Crank-Nicolson time scheme (CN)} \\ 1 & \text{Implicit backward time scheme (IT)} \end{cases} \tag{A.49}$$

In this work, the Crank-Nicolson scheme is used for time discretization. It combines the explicit and implicit scheme and is second-type. It is unconditionally stable. For further information, refer to Ref. [163].

The FV coefficients $a_{\mathrm{N}}, a_{\mathrm{S}}, a_{\mathrm{E}}, a_{\mathrm{W}}, a_{\mathrm{P}}^{0}, S_u, S_p$ are introduced in the following:

$$\begin{aligned}
a_{\mathrm{N}} &= -C_2 r_{\mathrm{P}} \Delta r \beta_{\mathrm{n}} + C_3 r_{\mathrm{P}} \Delta r \frac{1}{x_{\mathrm{PN}}} \\
a_{\mathrm{S}} &= C_2 r_{\mathrm{P}} \Delta r \beta_{\mathrm{s}} + C_3 r_{\mathrm{P}} \Delta r \frac{1}{x_{\mathrm{SP}}} \\
a_{\mathrm{E}} &= C_4 \Delta x \frac{r_{\mathrm{e}}}{r_{\mathrm{PE}}} \\
a_{\mathrm{W}} &= C_4 \Delta x \frac{r_{\mathrm{w}}}{r_{\mathrm{WP}}} \\
a_{\mathrm{P}}^{0} &= C_1 \frac{\Delta x r_{\mathrm{P}} \Delta r}{\Delta t} \\
S_u &= C_5 r_{\mathrm{P}} \Delta r \Delta x T_{\mathrm{s,P}}^{j} \\
S_p &= -C_5 r_{\mathrm{P}} \Delta r \Delta x
\end{aligned} \tag{A.50}$$

With these coefficients the equation simplifies to:

$$\begin{aligned}
& T_{\mathrm{P}}^{j+1} \left(a_{\mathrm{P}}^{0} + \Theta \left(a_{\mathrm{N}} + a_{\mathrm{S}} + a_{\mathrm{E}} + a_{\mathrm{W}} - S_p \right) \right) \\
& = \Theta \left(a_{\mathrm{N}} T_{\mathrm{N}}^{j+1} + a_{\mathrm{S}} T_{\mathrm{S}}^{j+1} + a_{\mathrm{E}} T_{\mathrm{E}}^{j+1} + a_{\mathrm{W}} T_{\mathrm{W}}^{j+1} \right) \\
& + (1-\Theta) \left(a_{\mathrm{N}} T_{\mathrm{N}}^{j} + a_{\mathrm{S}} T_{\mathrm{S}}^{j} + a_{\mathrm{E}} T_{\mathrm{E}}^{j} + a_{\mathrm{W}} T_{\mathrm{W}}^{j} \right) \\
& + T_{\mathrm{P}}^{j} \left(a_{\mathrm{P}}^{0} - (1-\Theta) \left(a_{\mathrm{N}} + a_{\mathrm{S}} + a_{\mathrm{E}} + a_{\mathrm{W}} - S_p \right) \right) + S_u
\end{aligned} \tag{A.51}$$

In order to solve the equation in MATLAB, the fluid energy equation is rearranged into a linear system of the type $Ax = b$ with A being a sparse matrix, T the vector for the unknown temperature and b the vector of the known figures.

$$\begin{aligned}
a_{\mathrm{P}} &= a_{\mathrm{P}}^{0} + \Theta \left(a_{\mathrm{N}} + a_{\mathrm{S}} + a_{\mathrm{E}} + a_{\mathrm{W}} - S_p \right) \\
A &= a_{\mathrm{P}} - \Theta \left(a_{\mathrm{N}} T_{\mathrm{N}}^{j+1} + a_{\mathrm{S}} T_{\mathrm{S}}^{j+1} + a_{\mathrm{E}} T_{\mathrm{E}}^{j+1} + a_{\mathrm{W}} T_{\mathrm{W}}^{j+1} \right) \\
b &= (1-\Theta) \left(a_{\mathrm{N}} T_{\mathrm{N}}^{j} + a_{\mathrm{S}} T_{\mathrm{S}}^{j} + a_{\mathrm{E}} T_{\mathrm{E}}^{j} + a_{\mathrm{W}} T_{\mathrm{W}}^{j} \right) \\
&+ T_{\mathrm{P}}^{j} \left(a_{\mathrm{P}}^{0} - (1-\Theta) \left(a_{\mathrm{N}} + a_{\mathrm{S}} + a_{\mathrm{E}} + a_{\mathrm{W}} - S_p \right) \right) + S_u
\end{aligned} \tag{A.52}$$

The equation $A T^{j+1} = b$ is solved by using the “backslash” operator in MATLAB: $T^{j+1} = A \backslash b$.

The energy equation inside a sperical particle is solved analogously. Equation 5.2 is integrated over volume and time. The radial variable is directed from “west” (core) to “east” (surface).

$$\begin{aligned}
& \int_0^{2\pi} \int_0^{\pi} \int_{y_{\mathrm{w}}}^{y_{\mathrm{e}}} \int_t^{t+\Delta t} \left[\rho_{\mathrm{s}} c_{\mathrm{s}} \frac{\partial T_{\mathrm{p}}}{\partial t} \right] dt \, y^2 dy \sin\phi d\phi d\varphi \\
& = \int_0^{2\pi} \int_0^{\pi} \int_{y_{\mathrm{w}}}^{y_{\mathrm{e}}} \int_t^{t+\Delta t} \left[\lambda_{\mathrm{s}} \left(\frac{\partial^2 T_{\mathrm{p}}}{\partial y^2} + \frac{2}{y} \frac{\partial T_{\mathrm{p}}}{\partial y} \right) \right] dt \, y^2 dy \sin\phi d\phi d\varphi
\end{aligned} \tag{A.53}$$

Again, the coefficients are shortened to C_9–C_{10} for simplification. The terms of the particle energy equations are integrated separately in the following equations:

$$\begin{aligned}&\int_0^{2\pi}\int_0^{\pi}\int_{y_\mathrm{w}}^{y_\mathrm{e}}\int_t^{t+\Delta t}\left[C_9\frac{\partial T_\mathrm{p}}{\partial t}\right]dt\,y^2dy\,\sin\phi\,d\phi\,d\varphi\\&\approx C_9 4\pi\,y_\mathrm{P}^2\Delta y(T_\mathrm{p,P}^{j+1}-T_\mathrm{p,P}^{j})\end{aligned}\tag{A.54}$$

$$\begin{aligned}&\int_0^{2\pi}\int_0^{\pi}\int_{y_\mathrm{w}}^{y_\mathrm{e}}\int_t^{t+\Delta t}\left[C_{10}\left(\frac{\partial^2 T_\mathrm{p}}{\partial y^2}+\frac{2}{y}\frac{\partial T_\mathrm{p}}{\partial y}\right)\right]dt\,y^2dy\,\sin\phi\,d\phi\,d\varphi\\&\approx C_{10}4\pi\Delta t\left[\Theta\left(y_\mathrm{e}^2\frac{T_\mathrm{p,E}}{\Delta y_\mathrm{PE}}+y_\mathrm{w}^2\frac{T_\mathrm{p,W}}{\Delta y_\mathrm{WP}}-T_\mathrm{p,P}\left(\frac{y_\mathrm{e}^2}{\Delta r_\mathrm{PE}}+\frac{y_\mathrm{w}^2}{\Delta y_\mathrm{WP}}\right)\right)^{j+1}\right.\\&\left.+(1-\Theta)\left(y_\mathrm{e}^2\frac{T_\mathrm{p,E}}{\Delta y_\mathrm{PE}}+y_\mathrm{w}^2\frac{T_\mathrm{p,W}}{\Delta y_\mathrm{WP}}-T_\mathrm{p,P}\left(\frac{y_\mathrm{e}^2}{\Delta r_\mathrm{PE}}+\frac{y_\mathrm{w}^2}{\Delta y_\mathrm{WP}}\right)\right)^{j}\right]\end{aligned}\tag{A.55}$$

Inserting the two terms back into Equation A.53 and dividing by $4\pi\Delta t$ leads to:

$$\begin{aligned}&C_9\frac{y_\mathrm{P}^2\Delta y}{\Delta t}(T_\mathrm{p,P}^{j+1}-T_\mathrm{p,P}^{j})\\&=C_{10}\left[\Theta\left(y_\mathrm{e}^2\frac{T_\mathrm{p,E}}{\Delta y_\mathrm{PE}}+y_\mathrm{w}^2\frac{T_\mathrm{p,W}}{\Delta y_\mathrm{WP}}-T_\mathrm{p,P}\left(\frac{y_\mathrm{e}^2}{\Delta r_\mathrm{PE}}+\frac{y_\mathrm{w}^2}{\Delta y_\mathrm{WP}}\right)\right)^{j+1}\right.\\&\left.+(1-\Theta)\left(y_\mathrm{e}^2\frac{T_\mathrm{p,E}}{\Delta y_\mathrm{PE}}+y_\mathrm{w}^2\frac{T_\mathrm{p,W}}{\Delta y_\mathrm{WP}}-T_\mathrm{p,P}\left(\frac{y_\mathrm{e}^2}{\Delta r_\mathrm{PE}}+\frac{y_\mathrm{w}^2}{\Delta y_\mathrm{WP}}\right)\right)^{j}\right]\end{aligned}\tag{A.56}$$

The FV coefficients for the particle energy equation are defined in the following:

$$\begin{aligned}a_\mathrm{E}&=C_{10}\frac{y_\mathrm{e}^2}{y_\mathrm{PE}}\\a_\mathrm{W}&=C_{10}\frac{y_\mathrm{w}^2}{y_\mathrm{WP}}\\a_\mathrm{P}^0&=C_9\frac{\Delta y_\mathrm{P}^2}{\Delta t}\\S_u&=S_p=0\end{aligned}\tag{A.57}$$

Inserting these coefficients in Equation A.56 leads to:

$$\begin{aligned}&T_\mathrm{p,P}^{j+1}\left(a_\mathrm{P}^0+\Theta\left(a_\mathrm{E}+a_\mathrm{W}-S_p\right)\right)\\&=\Theta\left(a_\mathrm{E}T_\mathrm{p,E}^{j+1}+a_\mathrm{W}T_\mathrm{p,W}^{j+1}\right)\\&+(1-\Theta)\left(a_\mathrm{E}T_\mathrm{p,E}^{j}+a_\mathrm{W}T_\mathrm{p,W}^{j}\right)\\&+T_\mathrm{p,P}^{j}\left(a_\mathrm{P}^0-(1-\Theta)\left(a_\mathrm{E}+a_\mathrm{W}-S_p\right)\right)+S_u\end{aligned}\tag{A.58}$$

The particle energy equation is solved analogously to the fluid energy equation using the "backslash" operator in MATLAB: $T_p^{j+1} = A\backslash b$. The boundary nodes are solved analogously, but including the boundary conditions by using ghost cells.

If a lumped capacitance is assumed (for Bi < 1), the energy equation of the solid phase can be solved analytically. The derivation of the analytical solution for the lumped capacitance model (Bi < 1) is shown in the following:

$$\begin{aligned} &\int_0^{2\pi}\int_{x_s}^{x_n}\int_{r_w}^{r_e}\int_t^{t+\Delta t}\left[(1-\varepsilon)\rho_s c_s \frac{\partial T_s}{\partial t}\right] dt\, r dr\, dx\, d\varphi \\ &= \int_0^{2\pi}\int_{x_s}^{x_n}\int_{r_w}^{r_e}\int_t^{t+\Delta t}[h_v(T-T_s)]\, dt\, r dr\, dx\, d\varphi \end{aligned} \tag{A.59}$$

The coefficients are simplified by using C_7 and C_8 in the following:

$$C_7\int_t^{t+\Delta t}\frac{\partial T_s}{(T-T_s)} = C_8\int_t^{t+\Delta t} dt \tag{A.60}$$

The discretized equation is then:

$$T_s^j = -\left(T^{j-1}-T_s^{j-1}\right)e^{-\frac{C_8}{C_7}\Delta t} + T^{j-1} \tag{A.61}$$

A.5.6 MATLAB subfunctions

In Table A.8 the sub-functions of *2dlumped.m* and *2dintgrad.m* are further explained and their in- and outputs are listed.

A.5.7 Verification with analytical solutions

The comparison of analytical and numerical solutions is performed using the non-dimensional temperatures $\theta(t) = \frac{T(t)-T_\infty}{T(t=0)-T_\infty}$, a non-dimensional time (Fourier number) $\tau = \mathrm{Fo} = \frac{\alpha t}{L_c^2}$, a non-dimensional heat transfer coefficient (Biot number) $\mathrm{Bi} = \alpha L_c/\lambda$ and normalized distances $X = \frac{x}{H}$, $R = \frac{r}{D/2}$ and $Y = \frac{y}{d/2}$. The equation for the fluid is split up as shown in Table A.9. Each term and the energy equation for the particle are solved separately.

The heat transfer term in the fluid equation (Equation 5.1a) can be solved analytically in an explicit function. The accordance of numerical and analytical results is shown in Figure A.4.
The thermal conduction term in axial direction (Equation 5.1b) is solved exemplarily for Bi = 1.0 (convection boundary condition) with the analytical equation for $\tau > 0.2$ for a plane wall (using only the first term of the infinite series). The eigenvalue λ_1 and the coefficient A_1 are 0.8603 and 1.1191 (for Bi = 1.0), respectively. The initial temperature of the plane wall is $\theta = 1$, the outer temperature is defined as $\theta = 0$. The accordance of numerical and analytical solutions for $\tau > 0.2$

is shown in Figure A.5. For short times, the temperature in the centre is the initial temperature and at the outer surface it is calculated with the error function according to Ref. [215].

Table A.8: Subfunctions

Subfunction	Description	Input	Output
fv2Mesh.m	mesh design in tank axes	number of CVs, total length of mesh	control volume coordinates, distances and widths
fvP1DMesh.m	mesh design in particle axis	number of CVs, total length of mesh	control volume coordinates, distances and widths
coefficients.m	calculation of equation coefficients C_i and heat transfer coefficient between filler and fluid	fluid temperature, fluid name, storage parameters	energy equation coefficients C_i and heat transfer coefficient
myfvDiffUDS.m	calculation of FV coefficients a_i for the five-point finite volume stencil on a Cartesian mesh	mesh, equation coefficients C_i	FV coefficients a_{N}, a_{S}, a_{W}, a_{E}
myfvbc.m	modification of FV coefficients a_i due to boundary condition	mesh, boundary conditions, FV coefficients a_i, energy equation coefficients C_i, fluid temperatures	FV coefficients a_{N}, a_{S}, a_{W}, a_{E}, a_{p}, a_{p0} and vector b
fvAmatrix.m	creation of sparse penta-diagonal matrix A	FV coefficients a_{N}, a_{S}, a_{W}, a_{E}, a_{p}	matrix A
myfvpost.m	Post-processing	mesh, temperature vector	temperature matrix

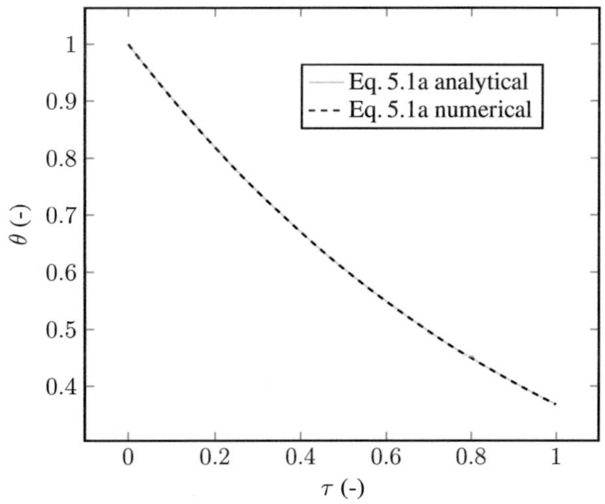

Figure A.4: Comparison of analytical and numerical solutions of fluid equation term 5.1a (heat transfer between fluid and solid)

Table A.9: Equation terms solved numerically and analytically

Equation	Term	Discretized form	Analytical solution
5.1a	$\varepsilon\rho_{\mathrm{f}}c_{\mathrm{pf}}\frac{\partial T}{\partial t} = -h_{\mathrm{v}}(T - T_{\mathrm{p}}\big\rvert_{y=\frac{d}{2}})$	$\varepsilon\rho_{\mathrm{f}}c_{\mathrm{pf}}\frac{r_{\mathrm{P}}\Delta r\Delta x}{\Delta t}(T_{\mathrm{P}}^{j+1} - T_{\mathrm{P}}^{j}) = -h_{\mathrm{v}}r_{\mathrm{P}}\Delta r\Delta x[\Theta T_{\mathrm{P}}^{j+1} +(1-\Theta)T_{\mathrm{P}}^{j}] +h_{\mathrm{v}}r_{\mathrm{P}}\Delta r\Delta x T_{\mathrm{p,P(y=d/2)}}^{j}$	$T(t) = e^{-h_{\mathrm{v}}/(\varepsilon\rho_{\mathrm{f}}c_{\mathrm{pf}})}(T(t = 0) - T_{\mathrm{p,P(y=d/2)}}) + T_{\mathrm{p,P(y=d/2)}}$ [156]
5.1b	$\varepsilon\rho_{\mathrm{f}}c_{\mathrm{pf}}\frac{\partial T}{\partial t} = k_{\mathrm{fx}}\frac{\partial^2 T}{\partial x^2}$	$\varepsilon\rho_{\mathrm{f}}c_{\mathrm{pf}}\frac{r_{\mathrm{P}}\Delta r\Delta x}{\Delta t}(T_{\mathrm{P}}^{j+1} - T_{\mathrm{P}}^{j}) = \varepsilon\lambda_{\mathrm{f}}r_{\mathrm{P}}\Delta r[\Theta(\frac{T_{\mathrm{N}}}{\Delta x_{\mathrm{PN}}} + \frac{T_{\mathrm{S}}}{\Delta x_{\mathrm{SP}}} -T_{\mathrm{P}}\left(\frac{1}{\Delta x_{\mathrm{PN}}} + \frac{1}{\Delta x_{\mathrm{SP}}}\right))^{j+1} +(1-\Theta)(\frac{T_{\mathrm{N}}}{\Delta x_{\mathrm{PN}}} + \frac{T_{\mathrm{S}}}{\Delta x_{\mathrm{SP}}} -T_{\mathrm{P}}\left(\frac{1}{\Delta x_{\mathrm{PN}}} + \frac{1}{\Delta x_{\mathrm{SP}}}\right))^{j}]$	$\theta(\tau,X) = \frac{T(t,x)-T_\infty}{T(t=0,x)-T_\infty} = A_1 e^{-\lambda_1^2\tau}\cos(\lambda_1 x/L)$, $(\tau > 0.2)$ [213]
5.1c	$\varepsilon\rho_{\mathrm{f}}c_{\mathrm{pf}}\frac{\partial T}{\partial t} = k_{\mathrm{fr}}\left(\frac{\partial^2 T}{\partial r^2} + \frac{1}{r}\frac{\partial T}{\partial r}\right)$	$\varepsilon\rho_{\mathrm{f}}c_{\mathrm{pf}}\frac{r_{\mathrm{P}}\Delta r\Delta x}{\Delta t}(T_{\mathrm{P}}^{j+1} - T_{\mathrm{P}}^{j}) = \varepsilon\lambda_{\mathrm{f}}\Delta x[\Theta(r_{\mathrm{e}}\frac{T_{\mathrm{E}}}{\Delta r_{\mathrm{PE}}} + r_{\mathrm{w}}\frac{T_{\mathrm{W}}}{\Delta x_{\mathrm{WP}}} -T_{\mathrm{P}}\left(\frac{r_{\mathrm{e}}}{\Delta r_{\mathrm{PE}}} + \frac{r_{\mathrm{w}}}{\Delta r_{\mathrm{WP}}}\right))^{j+1} +(1-\Theta)(r_{\mathrm{e}}\frac{{}_{\mathrm{E}}T}{\Delta r_{\mathrm{PE}}} + r_{\mathrm{w}}\frac{T_{\mathrm{W}}}{\Delta x_{\mathrm{WP}}} -T_{\mathrm{P}}\left(\frac{r_{\mathrm{e}}}{\Delta r_{\mathrm{PE}}} + \frac{r_{\mathrm{w}}}{\Delta r_{\mathrm{WP}}}\right))^{j}]$	$\theta(\tau,R) = \frac{T(t,r)-T_\infty}{T(t=0,r)-T_\infty} = A_1 e^{-\lambda_1^2\tau}J_0\lambda_1 R/r_0$, $(\tau > 0.2)$ [213]
5.1d	$\varepsilon\rho_{\mathrm{f}}c_{\mathrm{pf}}\frac{\partial T}{\partial t} = -\varepsilon\rho_{\mathrm{f}}c_{pf}u\frac{\partial T}{\partial x}$	$\varepsilon\rho_{\mathrm{f}}c_{\mathrm{pf}}\frac{r_{\mathrm{P}}\Delta r\Delta x}{\Delta t}(T_{\mathrm{P}}^{j+1} - T_{\mathrm{P}}^{j}) = -\varepsilon\rho_{\mathrm{f}}c_{pf}ur_{\mathrm{P}}\Delta r[\Theta(\beta_{\mathrm{n}}T_{\mathrm{N}} -\beta_{\mathrm{s}}T_{\mathrm{S}} + T_{\mathrm{P}}(\beta_{\mathrm{s}} - \beta_{\mathrm{s}}))^{j+1} +(1-\Theta)(\beta_{\mathrm{n}}T_{\mathrm{N}} - \beta_{\mathrm{s}}T_{\mathrm{S}} +T_{\mathrm{P}}(\beta_{\mathrm{s}} - \beta_{\mathrm{s}}))^{j}]$	$\theta(\tau,X) = (X - \tau)$ [214]
5.2	$\rho_{\mathrm{s}}c_{ps}\frac{\partial T_{\mathrm{p}}}{\partial t} = \lambda_{\mathrm{s}}\left(\frac{\partial^2 T_{\mathrm{p}}}{\partial y^2} + \frac{2}{y}\frac{\partial T_{\mathrm{p}}}{\partial y}\right)$	$\rho_{\mathrm{s}}c_{ps}\frac{y_{\mathrm{P}}^2\Delta y}{\Delta t}(T_{\mathrm{p,P}}^{j+1} - T_{\mathrm{p,P}}^{j}) = \lambda_{\mathrm{s}}[\Theta(y_{\mathrm{e}}^2\frac{T_{\mathrm{p,E}}}{\Delta y_{\mathrm{PE}}} + y_{\mathrm{w}}^2\frac{T_{\mathrm{p,W}}}{\Delta y_{\mathrm{WP}}} - T_{\mathrm{p,P}}(\frac{y_{\mathrm{e}}^2}{\Delta r_{\mathrm{PE}}} + \frac{y_{\mathrm{w}}^2}{\Delta y_{\mathrm{WP}}}))^{j+1} +(1-\Theta)(y_{\mathrm{e}}^2\frac{{}_{\mathrm{p,E}}T}{\Delta y_{\mathrm{PE}}} + y_{\mathrm{w}}^2\frac{T_{\mathrm{p,W}}}{\Delta y_{\mathrm{WP}}} - T_{\mathrm{p,P}}(\frac{y_{\mathrm{e}}^2}{\Delta r_{\mathrm{PE}}} + \frac{y_{\mathrm{w}}^2}{\Delta y_{\mathrm{WP}}}))^{j}]$	$\theta_{\mathrm{p}}(\tau,Y) = \frac{T_{\mathrm{p}}(t,y)-T_\infty}{T_{\mathrm{p}}(t=0,y)-T_\infty} = A_1 e^{-\lambda_1^2\tau}\frac{\sin(\lambda_1 r/r_0)}{\lambda_1 r/r_0)}$, $(\tau > 0.2)$ [213]

The thermal conduction in radial direction (Equation 5.1c) is solved analytically, as given in Table A.9, exemplary for Bi = 1.0 for a cylinder (convection boundary condition). The results are displayed in Figure A.6. Again, the equation is accurate only for $\tau > 0.2$. The coefficients of the analytical equation λ_1 and A_1 are 1.2558 and 1.2071 (for Bi = 1.0), respectively. The Bessel function of zeroth order is 1.0 for $r = 0$ and 0.6450 for $r = D/2$. The initial temperature of the cylinder is $\theta = 1$, the outer temperature is defined as $\theta = 0$. For short times, the temperature in the centre is the initial temperature and at the outer surface it is calculated with the error function according to Ref. [215].

The transport equation (Equation 5.1d) is solved analytically as given in Table A.9 for an initital condition $\Theta(\tau = 0, X) = X$ according to Ref. [214]. In Figure A.7 the numerical and analytical results are shown.

The intra-particle thermal diffusion (Equation 5.2) can be solved analytically, exemplarily demonstrated for a Biot number of 0.1 (molten salts) and Bi = 50 (molten metals) for a sphere (convection boundary condition), as shown in Figure A.8.

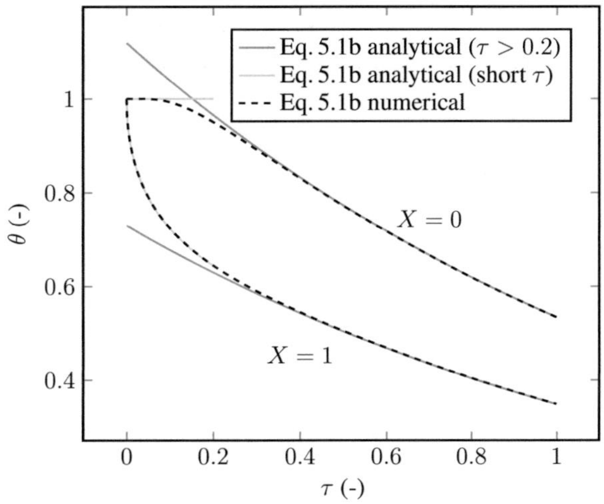

Figure A.5: Comparison of analytical and numerical solutions of energy equation term 5.1b (axial heat conduction with Bi = 1)

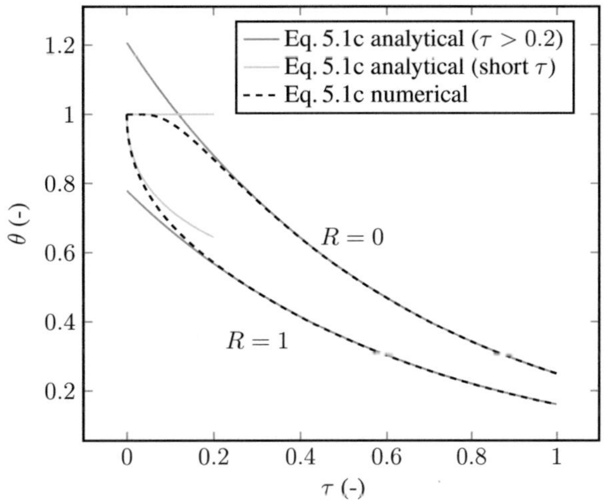

Figure A.6: Comparison of analytical and numerical solutions of energy equation term 5.1c (radial heat conduction with Bi = 1)

The analytical solution for $\tau > 0.2$ is solved with the eigenvalue λ_1 and coefficient A_1, which are 0.5423 and 1.0298 (for Bi = 0.1) and 3.0788 and 1.9962 (for Bi = 50) [213]. The initial temperature of the sphere is $\theta = 1$, the outer temperature is defined as $\theta = 0$. The numerical and analytical solutions show the same results for $\tau > 0.2$. For short times, the temperature in the centre is the initial temperature and at the outer surface it is calculated with the error function according to Ref. [215].

A.5.8 Validation with Promes-CNRS data with changed c_{ps}

In order to include the tank wall capacity, which is $\approx 23\%$ of the total capacity, the specific heat capacity of the solid filler material is adapted. The storage capacity of the filler material is artificially increased from Q_s = 340 kg·830 J/(kgK)·50 K = 3.9 kWh to $Q_s + Q_w = 5.8$ kWh to

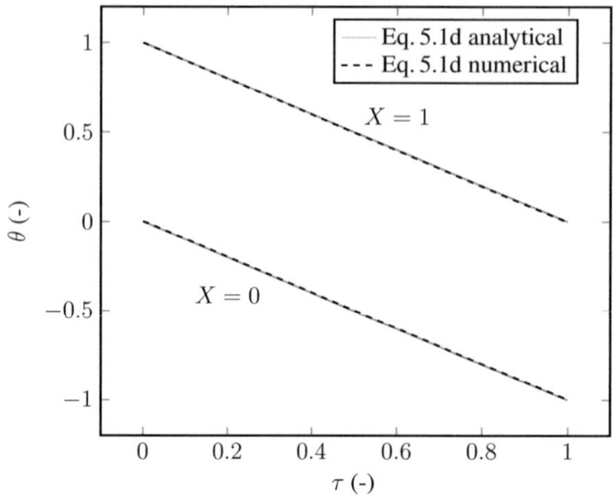

Figure A.7: Comparison of analytical and numerical solutions of energy equation term 5.1d (transport)

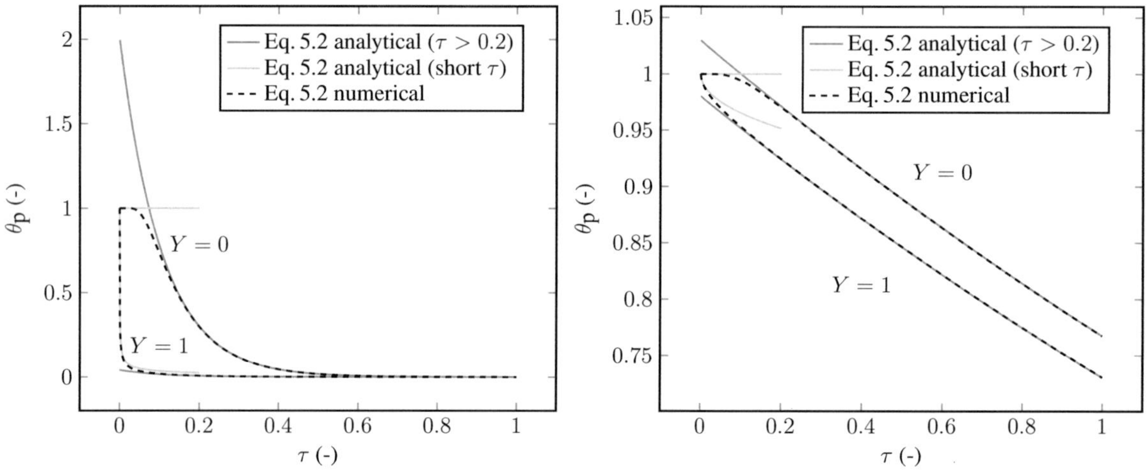

Figure A.8: Comparison of analytical and numerical solutions of Equation 5.2 (intra-particle thermal diffusion); left: for Bi = 50; right: for Bi = 0.1

represent the capacity of the tank wall in the filler material. This is implemented by adjusting the specific heat capacity of the filler material from 830 J/(kgK) to 1237 J/(kgK). Figure A.9 shows that the simulated fluid temperatures are now higher than before, as more thermal energy can be extracted from the solid material. However, now the simulation overestimates the experimental data. The deviation of the calculated from the experimental data is higher than in the case shown in Section 5.1.5, as shown in Table A.10.

Hoffmann et al. [113] solve the energy equation for the tank wall additionally and match the experimental values better. Therefore, it can be concluded that for a lab-scale storage tank, the tank wall capacity needs to be included in the simulation.

Table A.10: Deviation of modelled values from experimental data, *relative Mean Square Error

Error	Promes-CNRS 6.4 kWh$_{th}$	Promes-CNRS c_{ps} = 1237 J/(kgK)
Average (K)	2.62	5.00
Maximum (K)	10.83	26.23
rMSE* (10^{-4})	1.1	5.9

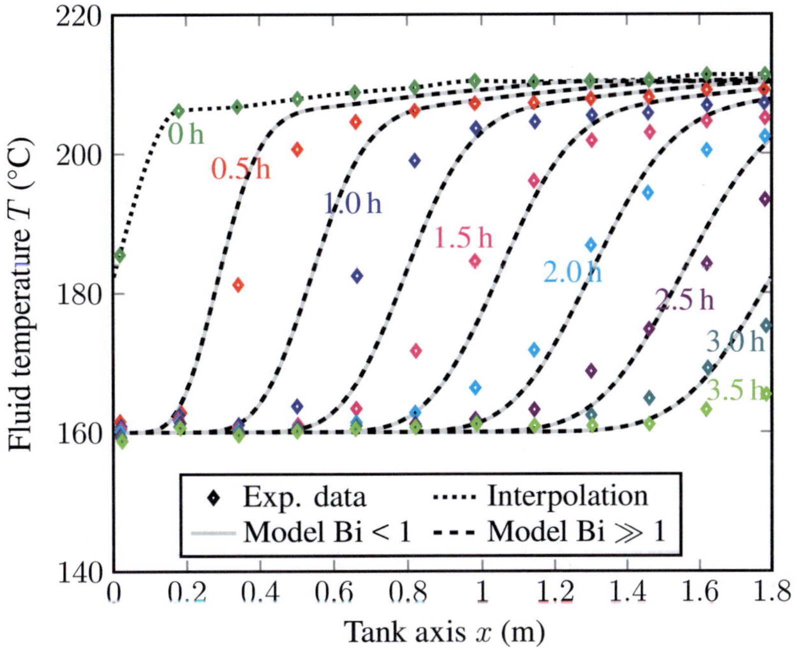

Figure A.9: Validation with experimental data from an experiment of Promes-CNRS (taken from Ref. [113]) with c_{ps} = 1237 J/(kgK)

A.5.9 Bidisperse packed bed

If particles of two different sizes are considered in the model, a second heat transfer term is added to the fluid energy equation:

$$\varepsilon \rho_f c_{pf} \left(\frac{\partial T}{\partial t} + u \frac{\partial T}{\partial x} \right) = k_{fx} \frac{\partial^2 T}{\partial x^2} - h_{v,1} (T - T_p|_{y=\frac{d_1}{2}}) - h_{v,2} (T - T_p|_{y=\frac{d_2}{2}}) \tag{A.62}$$

The specific surfaces $s_{v,i}$ to determine the volumetric heat transfer coefficients ($h_{v,i} = s_{v,i} \alpha_i$) are calculated by using the volumetric fraction of the particles of this size (Equation A.63) and the heat transfer coefficient α_i (Equation A.64), which is in the boundary condition at the particle surface.

The particle sizes and volume fraction of Pacheco et al. [39] are used; 30 wt% sand with diameter d_1 = 0.0015 mm and 70 wt% quartzite rocks with diameter d_2 = 0.01905 mm. Assuming the same density for the rocks and sand, this leads to specific surfaces of $s_{v,1}$ = 1200/m (sand) and $s_{v,2}$ = 220/m (rocks). The corresponding Sauter diameter [171] is $d_{32} = 6/\left(s_{v,1} + s_{v,2}\right) = 4.2\,\text{mm}$.

$$s_{v,i} = \frac{A_{p,tot,i}}{V_{tot}} = \frac{A_{p,tot,i}}{V_{p,tot,i}} \frac{V_{p,tot,i}}{V_{tot}} = \frac{\pi d_i^2}{\frac{1}{6}\pi d_i^3} \frac{V_{p,tot,i}}{V_{tot}} = \frac{6}{d_i} \frac{V_{p,tot,i}}{V_{tot}} \tag{A.63}$$

$$\alpha_i = \frac{\mathrm{Nu_{bed}} \lambda_f}{d_i} \tag{A.64}$$

In Figure A.10 the fluid temperatures along the tank axis in a packed bed with solar salt as heat transfer fluid are shown using two different diameters (bidisperse) and the Sauter diameter d_{32} (monodisperse). The resulting temperature distributions in a packed bed of 19-mm particles (monodisperse) and 1.5-mm particles (monodisperse) are added.

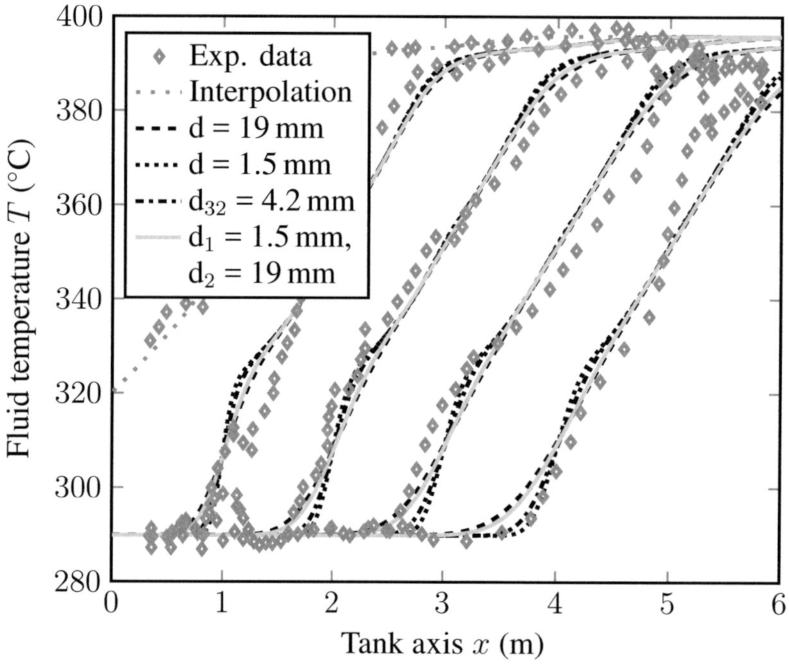

Figure A.10: Fluid temperature after 2 h and 4 h discharge time (first cycle) with different particle diameters in a monodisperse and a bisdisperse packed bed (d_1 = 1.5 mm and d_2 = 19 mm) in the first discharge cycle

A.6 Modelling of a packed bed storage with sodium – standby

A.6.1 Influence of k_{mix} models for solar salt

In Figure A.11 the temperature distributions in a solar salt/quartzite packed bed resulting from different models for k_{mix} are presented. In contrast to the results for liquid metal, the different models for k_{mix} do lead to similar temperature distributions in the packed bed due to the similar thermal conductivites of the fluid and solid phase.

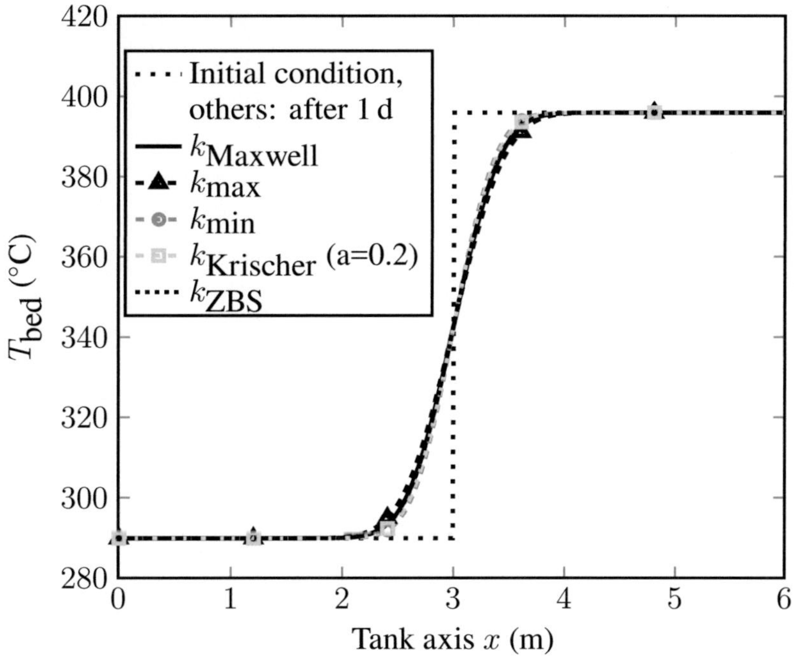

Figure A.11: Thermocline degradation of a solar salt/quartzite packed bed after 1 day of standby resulting from different k_{mix} (at T_{max}); $k_{Maxwell}$ = 1.73 W/(mK), k_{max} = 2.06 W/(mK), k_{min} = 1.36 W/(mK), $k_{Krischer}$ = 1.43 W/(mK), k_{ZBS} = 1.70 W/(mK)

A.6.2 Two-phase vs. one-phase model for solar salt

Figure A.12 shows the temperature distributions in a packed bed with solar salt after one day of standby resulting from a two-phase (heterogeneous) and one-phase (homogeneous) model. Similar results are obtained considering axial heat conduction in the solid phase in the heterogeneous model and considering a homogeneous model with a mixed effective conductivity and mixed thermal capacity. When neglecting the thermal conduction in axial direction in the heterogeneous model, the packed bed with solar salt experiences the least thermocline degradation, as now the relatively high thermal conductivity of the filler material is not included.

In Figure A.13 the influence of including heat losses in the models is presented for solar salt. Similar results are obtained from assuming an adiabatic and an insulated tank. However, using a heterogeneous model compared to a homogeneous models leads to large differences.

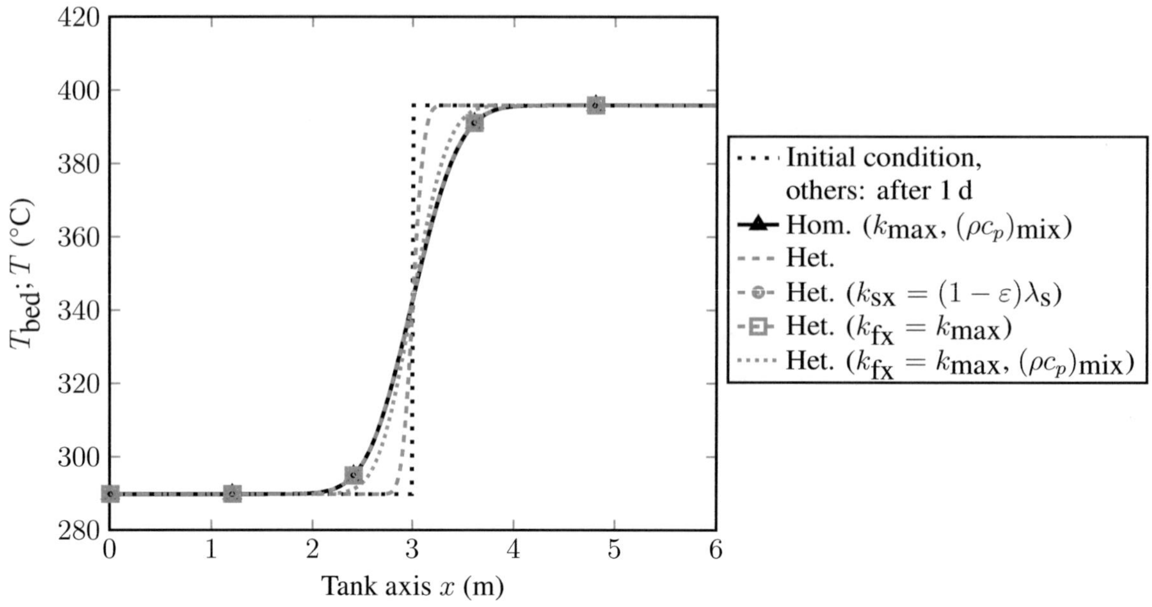

Figure A.12: Thermocline degradation of a solar salt/quartzite packed bed after 1 day of standby resulting from homogeneous and heterogeneous models

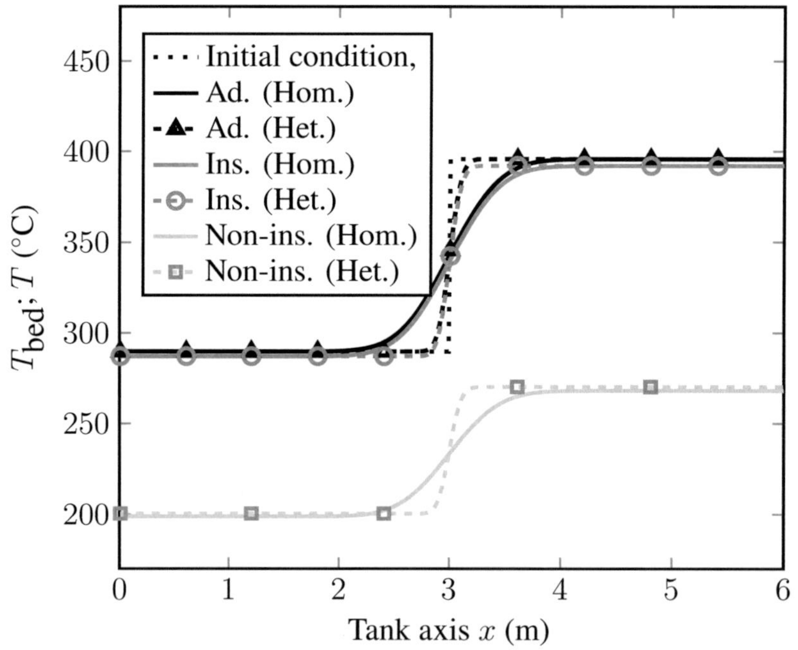

Figure A.13: Thermocline degradation of a solar salt/quartzite packed bed after 1 d of stand by resulting from different assumed heat losses, homogeneous (with k_{max}) and heterogeneous (without axial solid conduction)

A.6.3 Initial and boundary conditions

As initial condition, an ideal thermocline can be assumed:

$$t=0: \quad \begin{cases} T_{\text{mix}}(x,r)=T_{\text{min}} & 0 \leq x \leq H/2 \\ T_{\text{mix}}(x,r)=T_{\text{max}} & H/2 < x \leq H \end{cases} \tag{A.65}$$

Alternatively, if a discharge/charge step is combined with a subsequent standby period, the temperature distribution after the discharge/charge step can be used as initial condition:

$$t=0: \qquad T_{\text{mix}}(x,r)=f(x,r) \tag{A.66}$$

At the bottom and top of the tank no heat exchange is assumed:

$$x=0 \wedge x=H: \qquad \frac{\partial T_{\text{mix}}}{\partial x}=0 \tag{A.67}$$

If heat losses during standby are considered (in the 2D model), the boundary conditions at the center and tank wall are defined as follows:

$$r=0: \qquad \frac{\partial T_{\text{mix}}}{\partial r}=0 \tag{A.68}$$

$$r=R: \qquad -\lambda_f \cdot \frac{\partial T_{\text{mix}}}{\partial r}=h_{\text{w}}(T_{\text{mix}}-T_0) \tag{A.69}$$

A.6.4 Influence of heat losses for minimum D/H

For the minimum D/H in this parametric study, the highest heat losses are expected because of the largest wall surface. Therefore, it is investigated whether it is still tolerable to neglect heat losses and assume an adiabatic tank (Figure A.14).

The study is conducted as explained in Section 5.2.1. The initial condition is an ideal thermocline, the bottom half of the tank being at minimum temperature and the upper half at maximum temperature. The standby simulation is conducted with a one-dimensional one-phase model for 1 day for adiabatic conditions and for a well-insulated tank as described in Section 5.2.1. Figure A.14 shows that the insulated tank can still be assumed to be adiabatic accepting a 7 K temperature difference.

A.7 Mesh parameters

The mesh parameters Δx, Δr, Δy, Δt used for the simulations are given in the following tables.

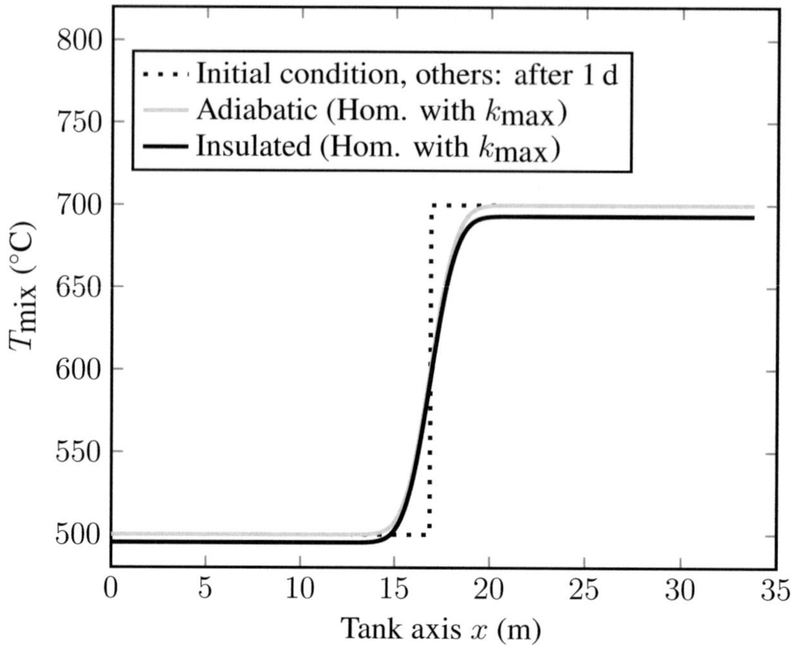

Figure A.14: Thermocline degradation of a sodium/quartzite packed bed with $D/H = 0.1$ after 1 day of standby resulting from different assumed heat losses

Table A.11: Mesh parameters for model comparisons – cycling

	Section 5.1.2		Section 5.1.3	
Case	(2P)1D	(2P)2D	$k_{f,i}$	Nu
nx	500	500	500	2000
Δx (m)	0.012	0.012	0.012	0.003
nr	–	15	–	–
Δr (m)	–	0.1	–	–
ny	50	40	50	70
Δy (mm)	0.7...0.004	0.8...0.013	0.7...0.004	0.8...0.0005
nt	5000	3000	5000	5000
Δt (s)	1.4	2.4	1.4	1.4

Table A.12: Mesh parameters for model comparisons – standby

	Section 5.2.1	Section 5.2.2		Section 5.2.3	
Case	$k_{\text{mix,i}}$	2P(1D)	1P(1D)	(1P)2D	(1P)1D
nx	500	500	500	500	500
Δx (m)	0.012	0.012	0.012	0.012	0.012
nr	–	–	–	30	–
Δr (m)	–	–	–	0.05	–
ny	–	30	–	–	–
Δy (mm)	–	0.7...0.0004	–	–	–
nt	5000	5000	5000	5000	5000
Δt (s)	1.4	1.4	1.4	1.4	1.4

Table A.13: Mesh parameters for the validation of discharge and standby simulation (Section 5.1.5 and 5.2.4) and mathematical verification (Section A.5.7)

	Discharge			Standby
	S.N.L.	Solar One	Promes-CNRS	S.N.L.
nx	600	500	500	500
Δx (m)	0.010	0.024	0.0036	0.012
ny	50	30	50	–
Δy (mm)	0.7...0.004	0.2...0.01	0.6...0.03	–
nt	4000	4000	3000	5000
Δt (s)	1.8	7.2	3.6	29.5
	Verification			
nX	300			
ΔX	0.003			
nY	300			
Δy	0.003			
$n\tau$	9000			
$\Delta\tau$	0.0001			

Table A.14: Mesh parameters for the parametric study (Section 6.3) (1/2)

ε	0.1	0.22	0.4	0.6	0.8	0.95
nx	2500	2500	2500	2500	2500	2500
Δx (m)	0.005	0.005	0.005	0.005	0.006	0.006
ny	70	70	70	70	70	70
Δy (mm)	0.8… 0.0005	0.8… 0.0005	0.8… 0.0005	0.8… 0.0005	0.8… 0.0005	0.8… 0.0005
nt	5000	5000	5000	5000	5000	5000
Δt (s)	2.9	2.9	2.9	2.9	2.9	2.9
D/H	0.1	0.3	0.5	1.0	1.5	2.0
nx	2500	2500	2500	2500	2500	2500
Δx (m)	0.014	0.006	0.005	0.003	0.002	0.002
ny	70	70	70	70	70	70
Δy (mm)	0.8… 0.0005	0.8… 0.0005	0.8… 0.0005	0.8… 0.0005	0.8… 0.0005	0.8… 0.0005
nt	5000	5000	5000	5000	5000	5000
Δt (s)	2.9	2.9	2.9	2.9	2.9	2.9
d (mm)	1.5 (mono)	19.0 (mono)	4.2 (mono)	1.5 (bi)	19.0 (bi)	
nx	1000	2500	2500	2000	2000	
Δx (m)	0.012	0.005	0.005	0.006	0.006	
ny	50	70	50	50	70	
Δy (mm)	0.07…0.0004	1.0…0.0007	0.2…0.001	0.07…0.0004	1.0…0.0007	
nt	20000	5000	15000	5000	5000	
Δt (s)	0.7	2.9	1.0	0.4	0.4	
d (mm)	1	5	15	30	50	
nx	1000	2500	2500	2500	2500	
Δx (m)	0.012	0.005	0.005	0.005	0.005	
ny	50	50	70	70	70	
Δy (mm)	0.05…0.0003	0.3…0.001	0.8…0.0005	1.5…0.001	2.5…0.002	
nt	30000	15000	5000	5000	5000	
Δt (s)	0.5	1.0	2.9	2.9	2.9	

Table A.15: Mesh parameters for the parametric study (Section 6.3) (2/2)

Filler	$Si0_2$	$MgAl_2O_4$	Al_20_3	Aust. steel	Fe
nx	2500	2500	2500	2500	2500
Δx (m)	0.005	0.005	0.004	0.004	0.004
ny	70	70	70	70	70
Δy (mm)	0.8...0.0005	0.8...0.0005	0.8...0.0005	0.8...0.0005	0.8...0.0005
nt	5000	5000	5000	5000	5000
Δt (s)	2.9	2.9	2.9	2.9	2.9
u_0 (mm/s)	0.5	1.0	1.9	3.5	5.0
nx	2500	2500	2500	2500	2500
Δx (m)	0.005	0.005	0.005	0.005	0.005
ny	70	70	70	70	70
Δy (mm)	0.8...0.0005	0.8...0.0005	0.8...0.0005	0.8...0.0005	0.8...0.0005
nt	20000	8000	5000	8000	20000
Δt (s)	2.7	3.4	2.9	1.0	0.3
T_{mean} (°C)	428	600	600		Opt. case
ΔT (K)	275	275	200		
nx	2500	2500	2500		2500
Δx (m)	0.004	0.004	0.005		0.008
ny	70	70	70		50
Δy (mm)	0.8...0.0005	0.8...0.0005	0.8...0.0005		0.3...0.001
nt	5000	5000	5000		20000
Δt (s)	2.9	2.9	2.9		0.7

Table A.16: Mesh parameters for the comparison with other fluids (Section 6.5)

Fluid (290°C - 565°C)	Na	LBE	Solar salt	
nx	2500	2500	500	
Δx (m)	0.004	0.004	0.02	
ny	70	70	–	
Δy (mm)	0.8...0.0005	0.8...0.0005	–	
nt	5000	5000	3000	
Δt (s)	2.9	2.9	4.8	
Fluid (500°C - 700°C)	Na	LBE	Pb	HTS
nx	2500	2500	2500	500
Δx (m)	0.005	0.005	0.005	0.02
ny	70	70	70	–
Δy (mm)	0.8...0.0005	0.8...0.0005	0.8...0.0005	–
nt	5000	5000	5000	3000
Δt (s)	2.9	2.9	2.9	4.8

Table A.17: Mesh parameters for the multi-tank arrangements (Section 7.3)

n_{tanks}	1	3	5	7
nx	2500	700	600	400
Δx (m)	0.005	0.011	0.011	0.015
ny	70	70	70	70
Δy (mm)	0.8...0.0005	0.8...0.0005	0.8...0.0005	0.8...0.0005
nt	5000	5000	8000	8000
Δt (s)	2.9	2.9	1.8	1.8
n_{tanks}	9	15	20	
nx	400	300	300	
Δx (m)	0.014	0.016	0.014	
ny	70	70	70	
Δy (mm)	0.8...0.0005	0.8...0.0005	0.8...0.0005	
nt	10000	12000	15000	
Δt (s)	1.4	1.2	1.0	
n_{tanks}		5		
D/H	0.1	0.5	2.0	
nx	700	600	200	
Δx (m)	0.028	0.011	0.013	
ny	70	70	70	
Δy (mm)	0.8...0.0005	0.8...0.0005	0.8...0.0005	
nt	8000	8000	12000	
Δt (s)	1.8	1.8	1.2	
n_{tanks}			20	
D/H	0.1	0.5	2.0	20.0
nx	300	300	100	200
Δx (m)	0.041	0.014	0.017	0.002
ny	70	70	70	70
Δy (mm)	0.8...0.0005	0.8...0.0005	0.8...0.0005	0.8...0.0005
nt	15000	15000	8000	5000
Δt (s)	1.0	1.0	1.8	2.9

Table A.18: Mesh parameters for the multi-tank arrangements in Gemasolar size (Section 7.3.4)

D/H			0.5	
n_{tanks}	1	3	5	7
nx	2500	1000	800	700
Δx (m)	0.015	0.026	0.027	0.028
ny	70	70	70	70
Δy (mm)	0.8...0.0005	0.8...0.0005	0.8...0.0005	0.8...0.0005
nt	10000	12000	15000	18000
Δt (s)	5.4	4.5	3.6	3.0
D/H	0.2	1.0		
n_{tanks}	20	1		
nx	400	1500		
Δx (m)	0.063	0.015		
ny	70	70		
Δy (mm)	0.8...0.0005	0.8...0.0005		
nt	20000	5000		
Δt (s)	2.7	10.8		

List of Figures

List of Tables

Bibliography

[1] R. Pitz-Paal, "Concentrating solar power: Still small but learning fast," *Nature Energy*, vol. 2, no. 7, p. 17095, 2017.

[2] International Energy Agency, "Renewables 2017 analysis and forecasts to 2022," tech. rep., OECD/IEA, 2017.

[3] U.S. Department of Energy, "Concentrated solar power," in *SunShot Vision Study*, ch. 5, pp. 97–123, U.S. Department of Energy, 2012.

[4] M. Mehos, C. Turchi, J. Vidal, M. Wagner, Z. Ma, C. Ho, W. Kolb, C. Andraka, and A. Kruizenga, "Concentrating solar power gen3 demonstration roadmap," Tech. Rep. NREL/TP-5500-67464, National Renewable Energy Laboratory (NREL), 2017.

[5] N. Boerema, G. Morrison, R. Taylor, and G. Rosengarten, "Liquid sodium versus Hitec as a heat transfer fluid in solar thermal central receiver systems," *Solar Energy*, vol. 86, no. 9, pp. 2293–2305, 2012.

[6] J. Pacio, A. Fritsch, C. Singer, and R. Uhlig, "Liquid metals as efficient coolants for high-intensity point-focus receivers: implications to the design and performance of next-generation CSP systems," *Energy Procedia*, vol. 49, no. 0, pp. 647–655, 2014. Proceedings of the SolarPACES 2013 International Conference.

[7] A. Fritsch, *Potenzialanalyse von Solarturmkraftwerken mit Flüssigmetallen als Wärmeträgermedium.* Logos Verlag Berlin, 2018.

[8] D. Stahl, F. Boese, and S. Kostrzewa, "System and components design of a sodium heat transfer circuit for solar power plants," *Electric Power Systems Research*, vol. 3, no. 3-4, pp. 151–161, 1980.

[9] J. Coventry, C. Andraka, J. Pye, M. Blanco, and J. Fisher, "A review of sodium receiver technologies for central receiver solar power plants," *Solar Energy*, vol. 122, pp. 749–762, 2015.

[10] B. D. Pomeroy, "Thermal energy storage in a packed bed of iron spheres with liquid sodium coolant," *Solar Energy*, vol. 23, pp. 513–515, 1979.

[11] J. Kotzé, *Thermal energy storage in metallic phase change materials.* PhD thesis, Stellenbosch University, 2014.

[12] G. C. Wilk, *Liquid metal based high temperature concentrated solar power: cost considerations.* PhD thesis, Georgia Institute of Technology, 2016.

[13] A. De la Calle, A. Sandoval, J. Hinkley, and J. Pye, "System-level simulation of a novel solar power tower plant based on a sodium receiver, PCM storage and sCO_2 power block," in *SolarPACES 2017*, 2017.

[14] Z. Kee, J. Coventry, and J. D. Pye, "A dynamic model of a sodium/salt PCM energy storage system," in *MATHMOD 2018 Extended Abstract Volume*, ARGESIM Publisher Vienna, 2018.

[15] International Energy Agency, "Technology roadmap energy storage," tech. rep., OECD/IEA, 2014.

[16] C. J. Winter and R. L. Sizmann, *Solar Power Plants: Fundamentals, Technology, Systems, Economics*. Springer-Verlag, 1991.

[17] International Energy Agency, "Technology roadmap solar thermal electricity," tech. rep., OECD/IEA, 2014.

[18] G. J. Kolb, D. J. Alpert, and C. W. Lopez, "Insights from the operation of solar one and their implications for future central receiver plants," *Solar Energy*, vol. 47, no. 1, pp. 39–47, 1991.

[19] J. E. Pacheco, H. E. Reilly, G. J. Kolb, and C. E. Tyner, "Summary of the Solar Two test and evaluation program," Tech. Rep. SAND2000-0372C, Sandia National Laboratories, 2000.

[20] U. Herrmann and D. W. Kearney, "Survey of thermal energy storage for parabolic trough power plants," *Journal of Solar Energy Engineering*, vol. 124, no. 2, pp. 145–152, 2002.

[21] R. I. Dunn, P. J. Hearps, and M. N. Wright, "Molten-salt power towers: Newly commercial concentrating solar storage," *Proceedings of the IEEE*, vol. 100, no. 2, pp. 504–515, 2012.

[22] NREL Concentrating Solar Power Projects Database: Power Tower Projects. `https://www.nrel.gov/csp/solarpaces/power_tower.cfm`. Accessed: 2018-08-28.

[23] HELIOSCSP - Solar Thermal Energy News. `http://helioscsp.com/`. Accessed: 2018-08-28.

[24] C. Kost, J. N. Mayer, J. Thomsen, N. Hartmann, C. Senkpiel, S. Philipps, S. Nold, S. Lude, and T. Schlegl, "Stromgestehungskosten Erneuerbare Energien," tech. rep., Fraunhofer ISE, 2013.

[25] International Renewable Energy Agency, "Renewables and electricity: A technology roadmap for REmap 2030," tech. rep., IRENA, 2015.

[26] O. Schmidt, A. Hawkes, A. Gambhir, and I. Staffell, "The future cost of electrical energy storage based on experience rates," *Nature Energy*, vol. 2, no. 8, p. 17110, 2017.

[27] A. Finkel, "Independent review into the future security of the national electricity market: Blueprint for the future," tech. rep., Commonwealth of Australia, 2017.

[28] Lazard, "Lazard's levelized cost of energy analysis - version 10.0," tech. rep., Lazard, 2016.

[29] D. Feldman, R. Margolis, and P. Denholm, "Exploring the potential competitiveness of utility-scale photovoltaics plus batteries with concentrating solar power, 2015-2030," Tech. Rep. NREL/TP-6A20-66592, National Renewable Energy Laboratory (NREL), 2016.

[30] M. Mehos, J. Jorgenson, P. Denholm, and C. Turchi, "An assessment of the net value of CSP systems integrated with thermal energy storage," *Energy Procedia*, vol. 69, pp. 2060–2071, 2015.

[31] European Solar Thermal Electricity Association, "Strategic research agenda 2020-2025," tech. rep., 2012.

[32] J. Lilliestam, M. Labordena, A. Patt, and S. Pfenninger, "Empirically observed learning rates for concentrating solar power and their responses to regime change," *Nature Energy*, vol. 2, no. 7, p. 17094, 2017.

[33] International Renewable Energy Agency, "Renewable power generation cost in 2017," tech. rep., IRENA, 2018.

[34] B. Kelly and D. Kearney, "Thermal storage commercial plant design study for a 2-tank indirect molten salt system - final report," Tech. Rep. NREL/SR-550-40166, National Renewable Energy Laboratory, 2006.

[35] W.-D. Steinmann, "21 - Thermal energy storage systems for concentrating solar power (CSP) technology," in *Advances in Thermal Energy Storage Systems* (L. F. Cabeza, ed.), Woodhead Publishing Series in Energy, pp. 511–531, Woodhead Publishing, 2015.

[36] P. A. Querol P, Olano J and V. T., "Single tank thermal storage prototype," *Proc. SolarPACES*, 2012.

[37] W. Hering, R. Stieglitz, and T. Wetzel, "Application of liquid metals for solar energy systems," *EPJ Web of Conferences*, vol. 33, p. 03003, 2012.

[38] S. Faas, "10 MWe solar thermal central receiver pilot plant: Thermal storage subsystem evaluation - final report," Tech. Rep. SAND86-8212, Sandia National Laboratories, Albuquerque, New Mexico, 1986.

[39] J. E. Pacheco, S. K. Showalter, and W. J. Kolb, "Development of a molten-salt thermocline thermal storage system for parabolic trough plants," *Journal of Solar Energy Engineering*, vol. 124, no. 2, pp. 153–159, 2002.

[40] Z. Ma, G. Glatzmaier, and M. Mehos, "Development of solid particle thermal energy storage for concentrating solar power plants that use fluidized bed technology," *Energy Procedia*, vol. 49, pp. 898–907, 2014.

[41] T. Baumann and S. Zunft, "Development and performance assessment of a moving bed heat exchanger for solar central receiver power plants," *Energy Procedia*, vol. 69, pp. 748–757, 2015.

[42] D. Laing and S. Zunft, "Using concrete and other solid storage media in thermal energy storage (TES) systems," in *Advances in Thermal Energy Storage Systems*, pp. 65–86, Elsevier, 2015.

[43] M. Jonemann, "Advanced thermal storage system with novel molten salt," Tech. Rep. NREL/SR-5200-58595, NREL, 2003.

[44] A. Fritsch, J. Flesch, V. Geza, C. Singer, R. Uhlig, and B. Hoffschmidt, "Conceptual study of central receiver systems with liquid metals as efficient heat transfer fluids," *Energy Procedia*, vol. 69, pp. 644–653, 2015.

[45] A. Gil, M. Medrano, I. Martorell, A. Lázaro, P. Dolado, B. Zalba, and L. F. Cabeza, "State of the art on high temperature thermal energy storage for power generation. Part 1 — concepts, materials and modellization," *Renewable and Sustainable Energy Reviews*, vol. 14, no. 1, pp. 31–55, 2010.

[46] B. Cardenas and N. Leon, "High temperature latent heat thermal energy storage: Phase change materials, design considerations and performance enhancement techniques," *Renewable and Sustainable Energy Reviews*, vol. 27, pp. 724–737, 2013.

[47] M. M. Kenisarin, "High-temperature phase change materials for thermal energy storage," *Renewable and Sustainable Energy Reviews*, vol. 14, no. 3, pp. 955–970, 2010.

[48] A. Sharma, V. Tyagi, C. Chen, and D. Buddhi, "Review on thermal energy storage with phase change materials and applications," *Renewable and Sustainable Energy Reviews*, vol. 13, no. 2, pp. 318–345, 2009.

[49] B. Zalba, J. M. Marin, L. F. Cabeza, and H. Mehling, "Review on thermal energy storage with phase change: materials, heat transfer analysis and applications," *Applied Thermal Engineering*, vol. 23, no. 3, pp. 251–283, 2003.

[50] J. P. Kotzé, "High temperature thermal energy storage utilizing metallic phase change materials and metallic heat transfer fluids," *Journal of Solar Energy Engineering*, vol. 135, p. 6, 2013.

[51] D. Laing, W.-D. Steinmann, R. Tamme, A. Wörner, and S. Zunft, "Advances in thermal energy storage development at the German Aerospace Center (DLR)," vol. 1, pp. 13–25, 2012.

[52] M. Johnson, J. Vogel, M. Hempel, B. Hachmann, and A. Dengel, "Design of high temperature thermal energy storage for high power levels," *Sustainable Cities and Society*, vol. 35, pp. 758–763, 2017.

[53] H. Zhang, J. Baeyens, G. Cáceres, J. Degrève, and Y. Lv, "Thermal energy storage: Recent developments and practical aspects," *Progress in Energy and Combustion Science*, vol. 53, pp. 1–40, 2016.

[54] P. Gimenez-Gavarrell and S. Fereres, "Glass encapsulated phase change materials for high temperature thermal energy storage," *Renewable Energy*, vol. 107, pp. 497–507, 2017.

[55] T. E. Alam, J. S. Dhau, D. Y. Goswami, and E. Stefanakos, "Macroencapsulation and characterization of phase change materials for latent heat thermal energy storage systems," *Applied Energy*, vol. 154, pp. 92–101, 2015.

[56] P. Galione, C. Peez-Segarra, I. Rodriguez, S. Torras, and J. Rigola, "Multi-layered solid-PCM thermocline thermal storage for CSP. Numerical evaluation of its application in a 50MWe plant," *Solar Energy*, vol. 119, pp. 134–150, 2015.

[57] V. Zipf, A. Neuhäuser, C. Bachelier, R. Leithner, and W. Platzer, "Assessment of different PCM storage configurations in a 50MWel CSP plant with screw heat exchangers in a combined sensible and latent storage – simulation results," *Energy Procedia*, vol. 69, pp. 1078–1088, 2015.

[58] H. Pointner, W.-D. Steinmann, and M. Eck, "Introduction of the PCM flux concept for latent heat storage," *Energy Procedia*, vol. 57, pp. 643–652, 2014.

[59] H. Pointner, W. Steinmann, M. Eck, and C. Bachelier, "Separation of power and capacity in latent heat energy storage," *Energy Procedia*, vol. 69, pp. 997–1005, 2015.

[60] E.-B. S. Mettawee and G. M. Assassa, "Thermal conductivity enhancement in a latent heat storage system," *Solar Energy*, vol. 81, no. 7, pp. 839–845, 2007.

[61] W. Zhao, D. M. France, W. Yu, T. Kim, and D. Singh, "Phase change material with graphite foam for applications in high-temperature latent heat storage systems of concentrated solar power plants," *Renewable Energy*, vol. 69, pp. 134–146, 2014.

[62] Mehling, "Latent heat storage using a PCM-graphite composite material," *TERRASTOCK 2000*, 2000.

[63] H. Michels and R. Pitz-Paal, "Cascaded latent heat storage for parabolic trough solar power plants," *Solar Energy*, vol. 81, no. 6, pp. 829–837, 2007.

[64] K. Elfeky, N. Ahmed, and Q. Wang, "Numerical comparison between single PCM and multi-stage PCM based high temperature thermal energy storage for CSP tower plants," *Applied Thermal Engineering*, vol. 139, pp. 609–622, 2018.

[65] P. Pardo, A. Deydier, Z. Anxionnaz-Minvielle, S. Rouge, M. Cabassud, and P. Cognet, "A review on high temperature thermochemical heat energy storage," *Renewable and Sustainable Energy Reviews*, vol. 32, pp. 591–610, 2014.

[66] J. Wu and X. F. Long, "Research progress of solar thermochemical energy storage," *International Journal of Energy Research*, vol. 39, no. 7, pp. 869–888, 2014.

[67] S. B. Viklund and M. T. Johansson, "Technologies for utilization of industrial excess heat: Potentials for energy recovery and CO2 emission reduction," *Energy Conversion and Management*, vol. 77, pp. 369–379, 2014.

[68] S. Brueckner, L. Miró, L. F. Cabeza, M. Pehnt, and E. Laevemann, "Methods to estimate the industrial waste heat potential of regions – a categorization and literature review," *Renewable and Sustainable Energy Reviews*, vol. 38, pp. 164–171, 2014.

[69] U.S. Department of Energy, "Waste heat recovery: Technology and opportiunities in U.S. industry," tech. rep., U.S. Department of Energy, 2008.

[70] S. Hirzel, B. Sonntag, and C. Rohde, "Kurzstudie Industrielle Abwärmenutzung," tech. rep., Fraunhofer ISI, 2013.

[71] A. Fernández, C. Barreneche, L. Miró, S. Brückner, and L. Cabeza, "Thermal energy storage (TES) systems using heat from waste," in *Advances in Thermal Energy Storage Systems*, pp. 479–492, Elsevier, 2015.

[72] A. Seitz, "Möglichkeiten thermischer Energiespeicher zur Flexibilisierung und Effizienzsteigerung energieintensiver Industrieprozesse," in *Energiespeichersymposium 2017*, 2017.

[73] M. Johnson, J. Vogel, M. Hempel, A. Dengel, M. Seitz, and B. Hachmann, "High temperature latent heat thermal energy storage integration in a co-gen plant," *Energy Procedia*, vol. 73, pp. 281–288, 2015.

[74] K. Niedermeier, J. Flesch, L. Marocco, and T. Wetzel, "Assessment of thermal energy storage options in a sodium-based CSP plant," *Applied Thermal Engineering*, vol. 107, pp. 386–397, 2016.

[75] C. Corgnale, B. Hardy, T. Motyka, R. Zidan, J. Teprovich, and B. Peters, "Screening analysis of metal hydride based thermal energy storage systems for concentrating solar power plants," *Renewable and Sustainable Energy Reviews*, vol. 38, pp. 821–833, 2014.

[76] P. A. Ward, C. Corgnale, J. A. Teprovich, T. Motyka, B. Hardy, B. Peters, and R. Zidan, "High performance metal hydride based thermal energy storage systems for concentrating solar power applications," *Journal of Alloys and Compounds*, vol. 645, pp. 374–378, 2015.

[77] M. Felderhoff and B. Bogdanovic, "High temperature metal hydrides as heat storage materials for solar and related applications," *International Journal of Molecular Sciences*, vol. 10, no. 1, pp. 325–344, 2009.

[78] M. Paskevicius, D. Sheppard, K. Williamson, and C. Buckley, "Metal hydride thermal heat storage prototype for concentrating solar thermal power," *Energy*, vol. 88, pp. 469–477, 2015.

[79] K. Lovegrove, A. Luzzi, I. Soldiani, and H. Kreetz, "Developing ammonia based thermochemical energy storage for dish power plants," *Solar Energy*, vol. 76, no. 1-3, pp. 331–337, 2004.

[80] C. Chen, H. Aryafar, K. M. Lovegrove, and A. S. Lavine, "Modeling of ammonia synthesis to produce supercritical steam for solar thermochemical energy storage," *Solar Energy*, vol. 155, pp. 363–371, 2017.

[81] M. Azpiazu, J. Morquillas, and A. Vazquez, "Heat recovery from a thermal energy storage based on the Ca(OH)2/CaO cycle," *Applied Thermal Engineering*, vol. 23, no. 6, pp. 733–741, 2003.

[82] M. Schmidt, C. Szczukowski, C. Rosskopf, M. Linder, and A. Wörner, "Experimental results of a 10 kW high temperature thermochemical storage reactor based on calcium hydroxide," *Applied Thermal Engineering*, vol. 62, no. 2, pp. 553–559, 2014.

[83] Y. Kato, M. Yamada, T. Kanie, and Y. Yoshizawa, "Calcium oxide/carbon dioxide reactivity in a packed bed reactor of a chemical heat pump for high-temperature gas reactors," *Nuclear Engineering and Design*, vol. 210, no. 1, pp. 1–8, 2001.

[84] M. Linder, "Using thermochemical reactions in thermal energy storage systems," *Advances in Thermal Energy Storage Systems*, pp. 357–374, 2015.

[85] Z. Yang and S. V. Garimella, "Thermal analysis of solar thermal energy storage in a molten-salt thermocline," *Solar Energy*, vol. 84, no. 6, pp. 974–985, 2010.

[86] Z. Yang and S. V. Garimella, "Cyclic operation of molten-salt thermal energy storage in thermoclines for solar power plants," *Applied Energy*, vol. 103, pp. 256–265, 2013.

[87] J. T. Van Lew, P. Li, C. L. Chan, W. Karaki, and J. Stephens, "Analysis of heat storage and delivery of a thermocline tank having solid filler material," *J. Sol. Energy Eng.*, vol. 133, no. 2, p. 021003, 2011.

[88] K. Reddy, V. Jawahar, S. Sivakumar, and T. Mallick, "Performance investigation of single-tank thermocline storage systems for CSP plants," *Solar Energy*, vol. 144, pp. 740–749, 2017.

[89] C. Xu, Z. Wang, Y. He, X. Li, and F. Bai, "Parametric study and standby behavior of a packed-bed molten salt thermocline thermal storage system," *Renewable Energy*, vol. 48, pp. 1–9, 2012.

[90] C. Xu, X. Li, Z. Wang, Y. He, and F. Bai, "Effects of solid particle properties on the thermal performance of a packed-bed molten-salt thermocline thermal storage system," *Applied Thermal Engineering*, vol. 57, no. 1-2, pp. 69–80, 2013.

[91] M. Cascetta, G. Cau, P. Puddu, and F. Serra, "Numerical investigation of a packed bed thermal energy storage system with different heat transfer fluids," *Energy Procedia*, vol. 45, pp. 598–607, 2014.

[92] A. Modi and C. D. Perez-Segarra, "Thermocline thermal storage systems for concentrated solar power plants: One-dimensional numerical model and comparative analysis," *Solar Energy*, vol. 100, pp. 84–93, 2014.

[93] A. G. Vilella and S. Yesilyurt, "Analysis of heat storage with a thermocline tank for concentrated solar plants: Application to AndaSol I," in *2015 IEEE International Conference on Industrial Technology (ICIT)*, IEEE, 2015.

[94] G. Angelini, A. Lucchini, and G. Manzolini, "Comparison of thermocline molten salt storage performances to commercial two-tank configuration," *Energy Procedia*, vol. 49, pp. 694–704, 2014.

[95] M. Biencinto, R. Bayón, E. Rojas, and L. González, "Simulation and assessment of operation strategies for solar thermal power plants with a thermocline storage tank," *Solar Energy*, vol. 103, pp. 456–472, 2014.

[96] G. J. Kolb, "Evaluation of annual performance of 2-tank and thermocline thermal storage systems for trough plants," *Journal of Solar Energy Engineering*, vol. 133, no. 3, p. 031023, 2011.

[97] D. Cocco and F. Serra, "Performance comparison of two-tank direct and thermocline thermal energy storage systems for 1 MWe class concentrating solar power plants," *Energy*, vol. 81, pp. 526–536, 2015.

[98] M. N. Strasser and R. P. Selvam, "A cost and performance comparison of packed bed and structured thermocline thermal energy storage systems," *Solar Energy*, vol. 108, pp. 390–402, 2014.

[99] C. Libby, "Solar thermocline storage systems: Preliminary design study," Tech. Rep. 1019581, Electric Power Research Institute EPRI, 2010.

[100] W. Gaggioli, F. Fabrizi, F. Fontana, L. Rinaldi, and P. Tarquini, "An innovative concept of a thermal energy storage system based on a single tank configuration using stratifying molten salts as both heat storage medium and heat transfer fluid, and with an integrated steam generator," *Energy Procedia*, vol. 49, pp. 780–789, 2014.

[101] S. Flueckiger, Z. Yang, and S. V. Garimella, "An integrated thermal and mechanical investigation of molten-salt thermocline energy storage," *Applied Energy*, vol. 88, no. 6, pp. 2098–2105, 2011.

[102] S. Flueckiger, Z. Yang, and S. V. Garimella, "Thermocline energy storage in the Solar One power plant: An experimentally validated thermomechanical investigation," in *ASME 2011 5th International Conference on Energy Sustainability, Parts A, B, and C*, ASME, 2011.

[103] I. González, O. Lehmkuhl, C. Pérez-Segarra, and A. Oliva, "Dynamic thermoelastic analysis of thermocline-like storage tanks," *Energy Procedia*, vol. 69, pp. 850–859, 2015.

[104] N. Sassine, F.-V. Donzé, A. Bruch, and B. Harthong, "Rock-bed thermocline storage: A numerical analysis of granular bed behavior and interaction with storage tank," Author(s), 2017.

[105] V. Dreißigacker, S. Zunft, and H. Müller-Steinhagen, "A thermo-mechanical model of packed-bed storage and experimental validation," *Applied Energy*, vol. 111, pp. 1120–1125, 2013.

[106] A. Mawire and M. McPherson, "Experimental and simulated temperature distribution of an oil-pebble bed thermal energy storage system with a variable heat source," *Applied Thermal Engineering*, vol. 29, no. 5-6, pp. 1086–1095, 2009.

[107] X. Yang, X. Yang, F. G. Qin, and R. Jiang, "Experimental investigation of a molten salt thermocline storage tank," *International Journal of Sustainable Energy*, vol. 35, no. 6, pp. 606–614, 2014.

[108] H. Yin, J. Ding, and X. Yang, "Experimental research on thermal characteristics of a hybrid thermocline heat storage system," *Applied Thermal Engineering*, vol. 62, no. 1, pp. 293–301, 2014.

[109] S. Rodat, A. Bruch, N. Dupassieux, and N. E. Mourchid, "Unique Fresnel demonstrator including ORC and thermocline direct thermal storage: Operating experience," *Energy Procedia*, vol. 69, pp. 1667–1675, 2015.

[110] A. Bruch, J. Fourmigue, and R. Couturier, "Experimental and numerical investigation of a pilot-scale thermal oil packed bed thermal storage system for CSP power plant," *Solar Energy*, vol. 105, pp. 116–125, 2014.

[111] A. Bruch, S. Molina, T. Esence, J. Fourmigué, and R. Couturier, "Experimental investigation of cycling behaviour of pilot-scale thermal oil packed-bed thermal storage system," *Renewable Energy*, vol. 103, pp. 277–285, 2017.

[112] T. Esence, R. Bayón, A. Bruch, and E. Rojas, "Study of thermocline development inside a dual-media storage tank at the beginning of dynamic processes," Author(s), 2017.

[113] J.-F. Hoffmann, T. Fasquelle, V. Goetz, and X. Py, "Experimental and numerical investigation of a thermocline thermal energy storage tank," *Applied Thermal Engineering*, vol. 114, pp. 896–904, 2017.

[114] T. Esence, A. Bruch, S. Molina, B. Stutz, and J.-F. Fourmigué, "A review on experience feedback and numerical modeling of packed-bed thermal energy storage systems," *Solar Energy*, vol. 153, pp. 628–654, 2017.

[115] C. Odenthal, N. Breidenbach, and T. Bauer, "Modelling and operation strategies of DLR's large scale thermocline test facility (TESIS)," *AIP Conference Proceedings*, vol. 1850, no. 1, p. 080019, 2017.

[116] C. Odenthal, F. Klasing, and T. Bauer, "Demonstrating cost effective thermal energy storage in molten salts: DLR's TESIS test facility," *Energy Procedia*, vol. 135, pp. 14–22, 2017.

[117] X. Yang, X. Yang, J. Ding, Y. Shao, F. G. Qin, and R. Jiang, "Criteria for performance improvement of a molten salt thermocline storage system," *Applied Thermal Engineering*, vol. 48, pp. 24–31, 2012.

[118] D. Brosseau, J. W. Kelton, D. Ray, M. Edgar, K. Chisman, and B. Emms, "Testing of thermocline filler materials and molten-salt heat transfer fluids for thermal energy storage systems in parabolic trough power plants," *Journal of Solar Energy Engineering*, vol. 127, no. 1, pp. 109–116, 2005.

[119] C. Martin, A. Bonk, M. Braun, C. Odenthal, and T. Bauer, "Investigation of the long-term stability of quartzite and basalt for a potential use as filler materials for a molten-salt based thermocline storage concept," *Solar Energy*, vol. 171, pp. 827–840, 2018.

[120] H. Grirate, N. Zari, A. Elmchaouri, S. Molina, and R. Couturier, "Stability testing of thermal oil in direct contact with rocks used as filler material for thermal energy storage in CSP power plants," *Energy Procedia*, vol. 69, pp. 860–867, 2015.

[121] H. Grirate, H. Agalit, N. Zari, A. Elmchaouri, S. Molina, and R. Couturier, "Experimental and numerical investigation of potential filler materials for thermal oil thermocline storage," *Solar Energy*, vol. 131, pp. 260–274, 2016.

[122] X. Py, N. Calvet, R. Olives, A. Meffre, P. Echegut, C. Bessada, E. Veron, and S. Ory, "Recycled material for sensible heat based thermal energy storage to be used in concentrated solar thermal power plants," *Journal of Solar Energy Engineering*, vol. 133, no. 3, p. 031008, 2011.

[123] N. Calvet, A. Meffre, J. C. Gomez, A. Faik, R. Olivès, X. Py, G. C. Glatzmaier, and S. Doppiu, "Post-industrial ceramics compatibility with heat transfer fluids for low-cost thermal energy storage applications in CSP," in *ASME 2012 6th International Conference on Energy Sustainability, Parts A and B*, ASME, 2012.

[124] F. Motte, Q. Falcoz, E. Veron, and X. Py, "Compatibility tests between solar salt and thermal storage ceramics from inorganic industrial wastes," *Applied Energy*, vol. 155, pp. 14–22, 2015.

[125] N. Calvet, G. Dejean, L. Unamunzaga, and X. Py, "Waste from metallurgic industry: A sustainable high-temperature thermal energy storage material for concentrated solar power," in *ASME 2013 7th International Conference on Energy Sustainability*, ASME, 2013.

[126] E. John, M. Hale, and P. Selvam, "Concrete as a thermal energy storage medium for thermocline solar energy storage systems," *Solar Energy*, vol. 96, pp. 194–204, 2013.

[127] K. M. Haneefa, M. Santhanam, and F. Parida, "Review of concrete performance at elevated temperature and hot sodium exposure applications in nuclear industry," *Nuclear Engineering and Design*, vol. 258, pp. 76–88, 2013.

[128] E. Miller, *Liquid Metals Handbook*, ch. 4 Corrosion of materials by liquid metals, pp. 144–183. Washington D.C.: Atomic Energy Commission, Department of the Navy, 1952.

[129] J. T. A. Roberts, "Analysis of potential attack on insulating materials exposed to sodium," Tech. Rep. NP-1986, Electric Power Research Institute EPRI, 1981.

[130] J. Stang, E. Simons, J. DeMastry, and J. Genco, "Compatibility of liquid and vapor alkali metals with construction materials," tech. rep., DTIC Document, 1966.

[131] P.-Y. Brisson, G. Soucy, M. Fafard, and M. Dionne, "The effect of sodium on the carbon lining of the aluminum electrolysis cell — a review," *Canadian Metallurgical Quarterly*, vol. 44, no. 2, pp. 265–280, 2005.

[132] G. Y. Lai, *High Temperature Corrosion and Materials Applications*. ASM International, 2007.

[133] A. W. Thorley and J. A. Bardsley, "Structural changes in materials exposed to liquid sodium," *Journal of the Royal Microscopical Society*, vol. 88, no. 4, pp. 431–447, 1968.

[134] J. H. Götze, ed., *Quartz, deposits, mineralogy and analytics*. Springer Geology, Berlin: Springer, 2012.

[135] C. Ullrich, T. Bodmer, C. Hübner, P. B. Kempa, E. Tsotsas, A. Eschner, G. Kasparek, F. Ochs, and M. H. Spitzner, "D6 Stoffwerte von Feststoffen," in *VDI-Wärmeatlas*, pp. 629–708, Springer Berlin Heidelberg, 2013.

[136] S. Kikuchi, N. Koga, H. Seino, and S. Ohno, "Kinetic study on liquid sodium–silica reaction for safety assessment of sodium-cooled fast reactor," *Journal of Thermal Analysis and Calorimetry*, vol. 121, no. 1, pp. 45–55, 2015.

[137] C. Casselman, "Consequences of interaction between sodium and concrete," *Nuclear Engineering and Design*, vol. 68, no. 2, pp. 207–212, 1982.

[138] H. R. Westrich, H. W. Stockmann, and A. Suo-Anttila, "Laboratory-scale sodium-carbonate aggregate concrete interactions," Tech. Rep. SAND83-1502, Sandia National Laboratories for the U.S. Department of Energy, Albuquerque, New Mexico, 1983.

[139] H. Bomelburg and C. Smith, *Sodium-NaK engineering handbook, Chapter 1 Physical Properties*. New York: Gordon & Breach, 1976.

[140] M. Bader and C. Busse, "Wetting by sodium at high temperatures in pure vapour atmosphere," *Journal of Nuclear Materials*, vol. 67, no. 3, pp. 295–300, 1977.

[141] M. Vanderhaegen and A. L. Belguet, "A review on sodium boiling phenomena in reactor systems," *Nuclear Science and Engineering*, vol. 176, no. 2, pp. 115–137, 2014.

[142] B. Sturm, U. Grigull, and H. Schmücker, "Strömungs- und Wärmeübergangsverhalten von flüssigem Quecksilber bei Benetzung und Nichtbenetzung," *Wärme- und Stoffübertragung*, vol. 19, no. 3, pp. 217–227, 1985.

[143] D. Reed, G. Coffey, E. Mast, N. Canfield, J. Mansurov, X. Lu, and V. Sprenkle, "Wetting of sodium on β"-Al_2O_3/YSZ composites for low temperature planar sodium-metal halide batteries," *Journal of Power Sources*, vol. 227, pp. 94–100, 2013.

[144] L. Viswanathan and A. V. Virkar, "Wetting characteristics of sodium on β"-alumina and on nasicon," *Journal of Materials Science*, vol. 17, no. 3, pp. 753–759, 1982.

[145] K. Ismail and R. Stuginsky Jr, "A parametric study on possible fixed bed models for pcm and sensible heat storage," *Applied Thermal Engineering*, vol. 19, no. 7, pp. 757–788, 1999.

[146] N. Wakao and S. Kaguei, *Heat and mass transfer in packed beds*. Topics in chemical engineering; 1, New York [u.a.]: Gordon and Breach, 1982.

[147] D. Kunii and J. M. Smith, "Heat transfer characteristics of porous rocks," *AIChE Journal*, vol. 6, no. 1, pp. 71–78, 1960.

[148] E.-U. Schlünder and E. Tsotsas, *Wärmeübertragung in Festbetten, durchmischten Schüttgütern und Wirbelschichten*. Lehrbuchreihe Chemieingenieurwesen/Verfahrenstechnik, Stuttgart: Thieme, 1988.

[149] J.-F. Hoffmann, T. Fasquelle, V. Goetz, and X. Py, "A thermocline thermal energy storage system with filler materials for concentrated solar power plants: Experimental data and numerical model sensitivity to different experimental tank scales," *Applied Thermal Engineering*, vol. 100, pp. 753–761, 2016.

[150] B. Alazmi and K. Vafai, "Analysis of variants within the porous media transport models," *Journal of Heat Transfer*, vol. 122, no. 2, p. 303, 2000.

[151] M. Fernández-Torrijos, C. Sobrino, and J. Almendros-Ibáñez, "Simplified model of a dual-media molten-salt thermocline tank with a multiple layer wall," *Solar Energy*, vol. 151, pp. 146–161, 2017.

[152] E. Tsotsas, "M7 Wärmeleitung und Dispersion in durchströmten Schüttungen," in *VDI-Wärmeatlas*, pp. 1517–1534, Springer Berlin Heidelberg, 2013.

[153] C. Xu, Z. Wang, Y. He, X. Li, and F. Bai, "Sensitivity analysis of the numerical study on the thermal performance of a packed-bed molten salt thermocline thermal storage system," *Applied Energy*, vol. 92, no. Supplement C, pp. 65–75, 2012.

[154] E. E. Gonzo, "Estimating correlations for the effective thermal conductivity of granular materials," *Chemical Engineering Journal*, vol. 90, no. 3, pp. 299–302, 2002.

[155] D. A. Nield and A. Bejan, *Convection in Porous Media*. Springer New York, 2013.

[156] F. Incropera, T. Bergman, D. DeWitt, and A. Lavine, *Fundamentals of Heat and Mass Transfer*. Wiley, 2013.

[157] B. Melissari and S. A. Argyropoulos, "Development of a heat transfer dimensionless correlation for spheres immersed in a wide range of Prandtl number fluids," *International Journal of Heat and Mass Transfer*, vol. 48, no. 21-22, pp. 4333–4341, 2005.

[158] V. Gnielinski, "G9 Wärmeübertragung Partikel – Fluid in durchströmten Haufwerken," in *VDI-Wärmeatlas*, pp. 839–840, Springer Berlin Heidelberg, 2013.

[159] P. von Böckh and T. Wetzel, *Wärmeübertragung*. Springer Berlin Heidelberg, 2014.

[160] V. Gnielinski, "G6 Querumströmte einzelne Rohre, Drähte und Profilzylinder," in *VDI-Wärmeatlas*, pp. 817–818, Springer Berlin Heidelberg, 2013.

[161] M. Kleiber and R. Joh, "D1 Berechnungsmethoden für Stoffeigenschaften," in *VDI-Wärmeatlas*, pp. 137–174, Springer Berlin Heidelberg, 2013.

[162] J. H. Ferziger and M. Perić, *Numerische Strömungsmechanik*. Springer Berlin Heidelberg, 2008.

[163] H. K. Versteeg and W. Malalasekera, *An introduction to computational fluid dynamics: the finite volume method*. Harlow [u.a.]: Pearson/Prentice Hall, 2. ed. ed., 2007.

[164] R. Bayón, E. Rivas, and E. Rojas, "Study of thermocline tank performance in dynamic processes and stand-by periods with an analytical function," *Energy Procedia*, vol. 49, pp. 725–734, 2014.

[165] M. Y. Haller, C. A. Cruickshank, W. Streicher, S. J. Harrison, E. Andersen, and S. Furbo, "Methods to determine stratification efficiency of thermal energy storage processes - review and theoretical comparison," *Solar Energy*, vol. 83, no. 10, pp. 1847–1860, 2009.

[166] M. Cascetta, G. Cau, P. Puddu, and F. Serra, "A comparison between CFD simulation and experimental investigation of a packed-bed thermal energy storage system," *Applied Thermal Engineering*, vol. 98, pp. 1263–1272, 2016.

[167] H. Bindra, P. Bueno, J. F. Morris, and R. Shinnar, "Thermal analysis and exergy evaluation of packed bed thermal storage systems," *Applied Thermal Engineering*, vol. 52, no. 2, pp. 255–263, 2013.

[168] S. A. Zavattoni, M. C. Barbato, A. Pedretti, and G. Zanganeh, "Single-tank TES system – transient evaluation of thermal stratification according to the second-law of thermodynamics," *Energy Procedia*, vol. 69, pp. 1068–1077, 2015.

[169] S. Ergun and A. A. Orning, "Fluid flow through randomly packed columns and fluidized beds," *Industrial & Engineering Chemistry*, vol. 41, no. 6, pp. 1179–1184, 1949.

[170] K. Allen, T. von Backström, and D. Kröger, "Packed bed pressure drop dependence on particle shape, size distribution, packing arrangement and roughness," *Powder Technology*, vol. 246, pp. 590–600, 2013.

[171] W. Kast, H. Nirschl, E. S. Gaddis, K.-E. Wirth, and J. Stichlmair, "L1 Einphasige Strömungen," in *VDI-Wärmeatlas*, pp. 1221–1284, Springer Berlin Heidelberg, 2013.

[172] M. Stieß, *Mechanische Verfahrenstechnik - Partikeltechnologie 1*. Springer Berlin Heidelberg, 3 ed., 2009.

[173] A. Koekemoer and A. Luckos, "Effect of material type and particle size distribution on pressure drop in packed beds of large particles: Extending the Ergun equation," *Fuel*, vol. 158, pp. 232–238, 2015.

[174] F. G. Casal, *Solar thermal power plants: achievements and lessons learned exemplified by the SSPS Project in Almeria/Spain*. Springer, 1987.

[175] J. Tombrink, "Untersuchung der Auswirkung von bi- und tridispersen Schüttungen auf den Wärmeübergang und den Druckverlust in thermoklinen Speichern mit Füllmaterial," Master's thesis, Leibniz Universität Hannover, 2017.

[176] K. Niedermeier, L. Marocco, J. Flesch, G. Mohan, J. Coventry, and T. Wetzel, "Performance of molten sodium vs. molten salts in a packed bed thermal energy storage," *Applied Thermal Engineering*, vol. 141, pp. 368–377, 2018.

[177] R. J. D. Tilley, *Understanding solids: the science of materials*. Chichester [u.a.]: Wiley, 2 ed., 2013.

[178] C. Odenthal, F. Klasing, and T. Bauer, "Parametric study of the thermocline filler concept based on exergy," *Journal of Energy Storage*, vol. 17, pp. 56–62, 2018.

[179] J. Flesch, A. Fritsch, G. Cammi, L. Marocco, F. Fellmoser, J. Pacio, and T. Wetzel, "Construction of a test facility for demonstration of a liquid lead-bismuth-cooled 10 kW thermal receiver in a solar furnace arrangement - SOMMER," *Energy Procedia*, vol. 69, pp. 1259–1268, 2015. International Conference on Concentrating Solar Power and Chemical Energy Systems, SolarPACES 2014.

[180] J. Osorio, A. Rivera-Alvarez, M. Swain, and J. Ordonez, "Exergy analysis of discharging multi-tank thermal energy storage systems with constant heat extraction," *Applied Energy*, vol. 154, pp. 333–343, 2015.

[181] R. M. Dickinson, C. A. Cruickshank, and S. J. Harrison, "Charge and discharge strategies for a multi-tank thermal energy storage," *Applied Energy*, vol. 109, pp. 366–373, 2013.

[182] D. Mather, K. Hollands, and J. Wright, "Single- and multi-tank energy storage for solar heating systems: fundamentals," *Solar Energy*, vol. 73, no. 1, pp. 3–13, 2002.

[183] W. Steinmann, C. Odenthal, and M. Eck, "System analysis and test loop design for the cellflux storage concept," *Energy Procedia*, vol. 49, pp. 1024–1033, 2014.

[184] D. Crandall and E. Thacher, "Segmented thermal storage," *Solar Energy*, vol. 77, no. 4, pp. 435–440, 2004.

[185] D. Schlipf, P. Schicktanz, H. Maier, and G. Schneider, "Using sand and other small grained materials as heat storage medium in a packed bed HTTESS," *Energy Procedia*, vol. 69, pp. 1029–1038, 2015.

[186] J. Rhee, A. Campbell, A. Mariadass, and B. Morhous, "Temperature stratification from thermal diodes in solar hot water storage tank," *Solar Energy*, vol. 84, no. 3, pp. 507–511, 2010.

[187] Solar Server. `https://www.solarserver.de/solar-magazin/nachrichten/archiv-2015/2015/kw22/abengoa-erhaelt-umweltgenehmigung-fuer-kombiniertes-photovoltaik-und-solarthermisches-kraftwerk-atacama-2-mit-210-mw-in-chile.html`. Accessed: 2018-08-28.

[188] Aussie Renewables. `https://www.aussierenewables.com.au/directory/jemalong-thermal-solar-station-266.html`. Accessed: 2018-08-28.

[189] H. Zhang, H. Wang, X. Zhu, Y.-J. Qiu, K. Li, R. Chen, and Q. Liao, "A review of waste heat recovery technologies towards molten slag in steel industry," *Applied Energy*, vol. 112, pp. 956–966, 2013.

[190] A. B. Zavoico, "Solar power tower design basis document," Tech. Rep. SAND2001-2100, Sandia National Laboratories, 2001.

[191] OECD-NEA, *Handbook on Lead-bismuth Eutectic Alloy and Lead Properties, Materials Compatibility, Thermal-hydraulics and Technologies*. OECD/NEA Nuclear Science Committee Working Party on Scientific Issues of the Fuel Cycle Working Group on Lead-bismuth Eutectic, 2007.

[192] OECD-NEA, *Handbook on Lead-bismuth Eutectic Alloy and Lead Properties, Materials Compatibility, Thermal-hydraulics and Technologies*. OECD/NEA Nuclear Science Committee Working Party on Scientific Issues of the Fuel Cycle Working Group on Lead-bismuth Eutectic, 2015.

[193] P. Li, E. Molina, K. Wang, X. Xu, G. Dehghani, A. Kohli, Q. Hao, M. H. Kassaee, S. M. Jeter, and A. S. Teja, "Thermal and transport properties of NaCl-KCl-$ZnCl_2$ eutectic salts for new generation high-temperature heat-transfer fluids," *Journal of Solar Energy Engineering*, vol. 138, no. 5, pp. 054501–1–8, 2016.

[194] Y. Li, X. Xu, X. Wang, P. Li, Q. Hao, and B. Xiao, "Survey and evaluation of equations for thermophysical properties of binary/ternary eutectic salts from NaCl, KCl, $MgCl_2$, $CaCl_2$, $ZnCl_2$ for heat transfer and thermal storage fluids in CSP," *Solar Energy*, vol. 152, pp. 57–79, 2017.

[195] D. F. Williams, "Assessment of candidate molten salt coolants for the NGNP/NHI heat-transfer loop," Tech. Rep. ORNL/TM-2006/69, Oak Ridge National Laboratory (ORNL), 2006.

[196] X. An, J. Cheng, P. Zhang, Z. Tang, and J. Wang, "Determination and evaluation of the thermophysical properties of an alkali carbonate eutectic molten salt," *Faraday Discuss.*, vol. 190, pp. 327–338, 2016.

[197] J. Pacio, C. Singer, T. Wetzel, and R. Uhlig, "Thermodynamic evaluation of liquid metals as heat transfer fluids in concentrated solar power plants," *Applied Thermal Engineering*, vol. 60, no. 1-2, pp. 295–302, 2013.

[198] Personal communication from SQM Europe M.V. with ANU. October 2015.

[199] Industrial Minerals. `http://www.indmin.com`. Accessed: 2017-06-15.

[200] USGS Mineral Yearbook, U.S. Geological Survey. `http://www.minerals.usgs.gov`. Accessed: 2017-02-15.

[201] Alibaba. `http://www.alibaba.com`. Accessed: 2016-12-15.

[202] Alibaba. `http://www.alibaba.com`. Accessed: 2018-11-08.

[203] P. Klein, T. Roos, and T. Sheer, "Parametric analysis of a high temperature packed bed thermal storage design for a solar gas turbine," *Solar Energy*, vol. 118, pp. 59–73, 2015.

[204] S. Hatte, C. Mira-Hernández, S. Advaith, A. Tinaikar, U. K. Chetia, K. Manu, K. Chattopadhyay, J. A. Weibel, S. V. Garimella, V. Srinivasan, and S. Basu, "Short and long-term sensitivity of lab-scale thermocline based thermal storage to flow disturbances," *Applied Thermal Engineering*, vol. 109, pp. 936–948, 2016.

[205] A. Andreozzi, B. Buonomo, O. Manca, P. Mesolella, and S. Tamburrino, "Numerical investigation on sensible thermal energy storage with porous media for high temperature solar systems," *Journal of Physics: Conference Series*, vol. 395, p. 012150, 2012.

[206] Y. Liu, Y. Liu, S. Tao, X. Liu, and Z. Wen, "Three-dimensional analysis of gas flow and heat transfer in a regenerator with alumina balls," *Applied Thermal Engineering*, vol. 69, no. 1-2, pp. 113–122, 2014.

[207] R. Bayón and E. Rojas, "Simulation of thermocline storage for solar thermal power plants: From dimensionless results to prototypes and real-size tanks," *International Journal of Heat and Mass Transfer*, vol. 60, pp. 713–721, 2013.

[208] T. Sanderson and G. Cunningham, "Performance and efficient design of packed bed thermal storage systems. Part 1," *Applied Energy*, vol. 50, no. 2, pp. 119–132, 1995.

[209] M. Wu, M. Li, C. Xu, Y. He, and W. Tao, "The impact of concrete structure on the thermal performance of the dual-media thermocline thermal storage tank using concrete as the solid medium," *Applied Energy*, vol. 113, pp. 1363–1371, 2014.

[210] A. Bonanos and E. Votyakov, "Sensitivity analysis for thermocline thermal storage tank design," *Renewable Energy*, vol. 99, pp. 764–771, 2016.

[211] B. Zhao, M. Cheng, C. Liu, and Z. Dai, "An efficient tank size estimation strategy for packed-bed thermocline thermal energy storage systems for concentrated solar power," *Solar Energy*, vol. 153, pp. 104–114, 2017.

[212] H. Klan and A. Thess, "F2 Wärmeübertragung durch freie Konvektion: Außenströmung," in *VDI Wärmeatlas*, pp. 757–764, Springer Berlin Heidelberg, 2013.

[213] Y. A. Cengel, *Introduction To Thermodynamics and Heat Transfer*. McGraw-Hill Science/Engineering/Math, 2007.

[214] G. W. Griffiths and W. E. Schiesser, *Traveling Wave Analysis of Partial Differential Equations: Numerical and Analytical Methods with Matlab and Maple*. Academic Press, 2012.

[215] H. Martin, "E2 Instationäre Wärmeleitung in ruhenden Körpern," in *VDI-Wärmeatlas*, pp. 727–752, Springer Berlin Heidelberg, 2013.

Bachelor and master theses supervised at IKET related to this work

Martin Pfuhl, Bachelor thesis: *Literaturrecherche zu thermischen Speicherkonzepten in konzentrierenden Solarkraftwerken*, 2015

Judith Kleinheins, Bachelor thesis: *Thermohydraulische numerische Untersuchungen in einem Festbett-Schichtenspeicher mit Natrium*, 2017

Niccola Zancan, Master thesis: *Numerical investigation of a sodium packed-bed thermal energy storage*, 2017

Manuel Meßmer, Master thesis: *Transiente Modellierung eines Systems für thermische Energiespeicher mit Flüssigmetallen als Wärmeträgermedien*, 2018

Publications within the framework of this work

K. Niedermeier, J. Flesch, L. Marocco and Th. Wetzel, *Assessment of thermal energy storage options in a sodium-based CSP plant*, Appl. Th. Eng. 2016 (107)

J. Flesch, K. Niedermeier, A. Fritsch, D. Musaeva, L. Marocco, R. Uhlig, E. Baake, R. Buck and T. Wetzel, *Liquid metals for solar power systems*, IOP Conf. Ser.: Mat. Sc. and Eng. 2017 (228)

K. Niedermeier, L. Marocco, J. Flesch, G. Mohan, J. Coventry and Th. Wetzel, *Performance of molten sodium vs. molten salts in a packed bed thermal energy storage*, Appl. Th. Eng. 2018 (141)

J. Flesch, L. Marocco, A. Fritsch, K. Niedermeier and Th. Wetzel, *Entropy generation minimization analysis of Solar Salt, sodium and lead-bismuth eutectic as high temperature heat transfer fluids*, submitted to ASME J. Heat Transfer

Presentations at conferences within the framework of this work

J. Flesch, K. Niedermeier, P. Dan, G. Cammi, L. Marocco, H. Piecha and Th. Wetzel, *Bestimmung von Wärmeübergangskoeffizienten mit einseitiger Randbedingung an einem mit Flüssigmetall durchströmten Rohr* (Poster), Jahrestreffen 2015 der Fachgruppen Trocknungstechnik und Wärmeübertragung, 4th–6th March 2015, Leipzig

L. Marocco, F. Fellmoser, J. Flesch, K. Niedermeier and T. Wetzel, *Numerical investigation of turbulent forced convection to liquid metal flowing inside the receiver tube of a solar tower* (Poster), SolarPACES conference, 13th–16th October 2015, Cape Town, South Africa

K. Niedermeier and Th. Wetzel, *Liquid metals as heat transfer fluids for high temperature processes* (Talk), Japanese-German Workshop on Renewable Energies, 1st–4th March 2016, Tokio, Japan

K. Niedermeier, J. Flesch, L. Marocco and Th. Wetzel, *Assessment of thermal energy storage options in a liquid metal-cooled receiver* (Poster), SolarPACES conference, 11th–14th October 2016, Abu Dhabi, UAE

K. Niedermeier, J. Kleinheins, J. Flesch, L. Marocco and Th. Wetzel, *Numerical investigation of a sodium dual-media thermal energy storage system for CSP* (Poster), 11th International Renewable Energy Storage Conference, 14th–16th March 2017, Düsseldorf

K. Niedermeier, J. Flesch, L. Marocco and Th. Wetzel, *Thermische Energiespeicherung für ein Natrium-gekühltes Solarturmkraftwerk* (Talk), 6. Jahrestagung des KIT-Zentrums Energie – Doktorandensymposium, 2nd June 2017, Karlsruhe

Presentations at meetings of the Liquid Metal Technologies (LIMTECH) Alliance

K. Niedermeier, J. Flesch, L. Marocco and Th. Wetzel, *Thermal energy storage for a LBE loop in a concentrating solar thermal system* (Poster), LIMTECH PhD summerschool, 30th–1st July 2015, Karlsruhe

K. Niedermeier, J. Flesch, F. Fellmoser, A. Jekel, K. Wittemann, L. Marocco and Th. Wetzel, *Thermal energy storage options for liquid metal based CSP* (Poster), LIMTECH Midterm evaluation, 30th September 2015, Dresden-Rossendorf

K. Niedermeier, J. Flesch, L. Marocco and Th. Wetzel, *Thermal energy storage for liquid metal-based CSP* (Poster), LIMTECH PhD Summer School, 12th–13th July 2016, Ilmenau

K. Niedermeier, J. Flesch, L. Marocco and Th. Wetzel, *Thermal energy storage for liquid metal-based CSP* (Poster), LIMTECH Jahrestreffen, 3rd–4th November 2016, Freiberg

J. Flesch, K. Niedermeier, A. Fritsch, D. Musaeva, L. Marocco, R. Uhlig, E. Baake, R. Buck and Th. Wetzel *Liquid metals for solar power systems* (Poster), LIMTECH-Symposium, 19th–20th September 2017, Dresden-Rossendorf